A Companion to Martin Heidegger's "Being and Time"

Current Continental Research
is co-published by
The Center for Advanced Research
in Phenomenology
and
University Press of America, Inc.

CURRENT CONTINENTAL RESEARCH 550

Joseph J. Kockelmans, Editor

A COMPANION TO MARTIN HEIDEGGER'S "BEING AND TIME"

1986

Center for Advanced Research in Phenomenology & University Press of America, Washington, D.C.

B
3279
.H49
C65
1986

Library of Congress Cataloging in Publication Data
Main entry under title:

A Companion to Martin Heidegger's "Being and time."

(Current continental research ; 550)
Bibliography: p.
Includes indexes.
Contents: Signification and radical subjectivity in
Heidegger's Habilitationsschrift / Roderick M. Stewart
— Heidegger's early lecture courses / Theodore J.
Kisiel — Heidegger's 'Introduction to the
phenomenology of religion', 1920-1921 / Thomas J.
Sheehan —[etc.]
    1. Heidegger, Martin, 1889-1976—Addresses, essays,
lectures.   2. Heidegger, Martin, 1889-1976.   Sein und
Zeit—Addresses, essays, lectures.   3. Ontology—
Addresses, essays, lectures.   4. Space and time—
Addresses, essays, lectures.   I. Kockelmans, Joseph J.,
1923-   II. Center for Advanced Research in
Phenomenology.   III. Series.
B3279.H49C65   1986   111   85-29569
ISBN 0-8191-5196-3 (alk. paper)
ISBN 0-8191-5197-1 (pbk. : alk. paper)

Table of Contents

iv

ACKNOWLEDGMENTS

Several essays contained in this anthology have appeared elsewhere. Grateful acknowledgment is made to the publishers, editors, and authors represented here for their generosity in granting permission to reprint selections from copyright material and for their cooperation throughout the work.

The other essays were expressly written for this volume. The copyright of these essays belongs to the authors, namely Theodore J. Kisiel, John Caputo, Joseph Kockelmans, Thomas Sheehan, Marion Heinz, and Graeme Nicholson. It goes without saying that I am particularly grateful for their cooperation in and contribution to this volume.

INTRODUCTION

There is common agreement among philosophers today that Heidegger's *Sein und Zeit* is one of the most important philosophical works written in the twentieth century. The work has been read and studied by thousands; it has been commented upon by a great number of scholars; finally, it has had an enormous influence, not only on philosophy itself, but also on theology, literary criticism, aesthetics, psychology, psychiatry, and the social sciences.

There is also common agreement among the "experts" that *Being and Time* is not an easy book to read. The reason for this appears to be that like Aristotle's *Metaphysics*, Kant's *Critique of Pure Reason*, and Hegel's *Phenomenology of Spirit*, it is a book that is quite different from most other philosophical works written during the same period, and in many respects is far ahead of its time. Even today the person who comes in contact with this important book for the first time will encounter a wealth of ideas which at first sight seem to run contrary to what one expected a philosophical treatise to be about. The book does not contain a metaphysics; nor does it have the character of an ethics or a political philosophy; it most certainly is not a philosophy of science in the common conception of this term, either. At first the book seems to be some kind of philosophical anthropology or "rational psychology." Yet the reader realizes soon that its author definitely and categorically denies this to be the case. Heidegger explicitly and repeatedly states that the book is meant to introduce the reader to a philosophical reflection on the question concerning the meaning of Being.

The book was supposed to have had two major parts, both subdivided into three major subdivisions. Yet in 1927 the work was published in an incomplete form, partly due to time pressures, partly as a consequence of philosophical problems which Heidegger had been unable to solve at that time. In its present form the book contains only the first two major subdivisions of the first part.

vii

In *Being and Time* Heidegger attempts to apply a "hermeneutic phenomenology" to an analytic of man's mode of Being, and carefully explains first the sense in which hermeneutic phenomenology is to be understood. In Heidegger's opinion the basic problem underlying philosophy's main concern is to be found in the question concerning the meaning of Being. This question is to be dealt with in *ontology*; yet such an ontology is to be prepared by a *fundamental ontology* which must take the form of an eksistential analytic of man's mode of Being to be understood as Being-in-the-world. In the author's view it is particularly in this fundamental ontology that the hermeneutic phenomenological method is to be employed. From the outset Heidegger makes it quite clear in *Being and Time* that what is to be understood by hermeneutic phenomenology is not identical with Husserl's transcendental phenomenology. He explicitly claims the right to develop the idea of phenomenology in his own way, beyond the stage to which it had been brought by Husserl himself. On the other hand, it is clear also that Heidegger sees in Husserl's phenomenology the indispensable foundation for such a further development. The reason for Heidegger's being unable to follow Husserl more closely is to be found in Husserl's conception of the transcendental reduction and his idea that the ultimate source of all meaning consists in transcendental subjectivity which as such originally is world-less. This explains why Heidegger tries to conceive man's Being as Being-in-the-world.

As the title of the book suggests the concept of time occupies an important place in *Being and Time*. Already in the brief preface to the book Heidegger indicates how Being and time are to be related. "Our aim in the following treatise is to work out the question concerning the meaning of Being... Our provisional aim is the interpretation of time as the possible horizon for any understanding whatsoever of Being."

In the first division of Part I Heidegger takes as his guiding clue the fact that the essence of man consists in his ek-sistence; that toward which man stands out is the world; thus one can also say that the essence of man is Being-in-the-world. The main task of this first division is to unveil the precise meaning of this compound expression; but in so doing the final goal remains the preparation of an answer for the question concerning the meaning of Being. Heidegger justifies this approach to the Being question by pointing out that man taken as Being-in-the-world is the only being who can make himself transparent in his own mode of Being. The very asking of this question is one of this being's

basic modes of Being; and as such it receives its essential character from what is inquired about, namely Being itself. "This
entity which each of us is himself and which includes inquiring as
one of the possibilities of its Being we shall denote by the term
'Dasein'." Thus the technical term "Dasein" which usually is left
untranslated, refers to man precisely insofar as he essentially
relates to Being.

   The preparatory analysis of Dasein's mode of Being can only
serve to *describe* the essence of this being; it cannot *interpret*
its meaning ontologically. The preparatory analysis merely tries to
lay bare the horizon for the most primordial way of interpreting
Being. Once this horizon has been made visible, the preparatory
analysis is to be replaced by a genuinely ontological interpretation. The horizon referred to here is temporality which thus
determines the meaning of the Being of Dasein. This is the reason
why all the structures of man's Being exhibited in the first
division are to be re-interpreted in the second as modes of temporality. But even in interpreting Dasein as temporality, the
question concerning the meaning of Being is not yet answered;
only the ground is prepared here for later obtaining such an
answer. *Being and Time* was thus meant to lay the foundations
for an ontology (metaphysics) and, with Kant, to stress the finitude of man in any attempt to found metaphysics.

   In *Being and Time* Heidegger uses the phenomenological
method. For him phenomenology *(legein ta phainomena:* to let what
shows itself be seen from itself) is that method by means of which
we let that which of its own accord manifests itself, reveal itself
as it is. The "thing itself" to be revealed in *Being and Time* is
man taken as Dasein. Thus *Being and Time* attempts to let Dasein
reveal itself in what and how it is, and the analysis shows concretely that the genuine self of Dasein consists in the process of
finite transcendence whose ultimate meaning is time.

   Characteristic for Dasein is its understanding of Being and
this is the process by which Dasein transcends beings in the
direction of Being, and comprehends all beings, itself included,
in their Being. This explains why the essence of Dasein can also
be defined as transcendence and freedom. It should be stressed
here at once that the process of transcendence is inherently
finite. For, first of all, Dasein is not master over its own origin;
it simply finds itself thrown among beings (thrownness). Secondly, thrown among beings, Dasein must concern itself with these
beings and, thus, has the tendency to lose itself among them
(fallenness), and to forget its ontological "destination." Finally,

transcendence is a process which inherently is unto Dasein's end, *death*. The ground of the negativity that manifests itself in these modalities is what Heidegger calls "guilt," which is here not to be understood in a moral sense.

The process of transcendence itself is rooted in the unity of *time*. It is through the finite process of transcendence that Dasein discloses Being in the beings. Transcendence can also be comprehended as the coming-to-pass of the truth. In the tradition truth was commonly understood as the conformity between intellect and thing, between judgment and being. This conception of truth has its merits and, thus, is to be maintained. Yet this conception is without foundation. Truth in the sense of conformity presupposes that the thing which manifests itself to the knower as that which is to be judged in a determinate way, is already discovered. The discovery of beings by Dasein is, therefore, a more radical form of truth than the truth taken as the conformity between judgment and thing. Now the process of finite transcendence precisely consists in this discovering, this letting beings be manifest. The finitude of the process of transcendence thus reflects itself in the coming-to-pass of the truth in Dasein's discovery. Every apprehension by Dasein is at the same time also a misapprehension; Dasein's uncovering is at the same time also a covering-up; revealment implies in each case some form of concealment; what is discovered is at the same time also partly hidden, and soon it may even slip back again into forgottenness. Also Dasein is equiprimordially both in the truth and in untruth.

The basic structure of finite transcendence consists of 1) understanding *(Verstehen)*, i.e., the component in and through which Dasein projects the world; 2) ontological disposition or moodness *(Befindlichkeit)*, i.e., the component through which Dasein's thrownness, fallenness, and the world's non-Being are disclosed; and 3) *logos (Rede)*, i.e., the component through which Dasein can unfold and articulate "in language" what understanding and original mood disclose. These components constitute a unity insofar as transcendence essentially is care *(Sorge)*: ahead of itself, being already in the world, as being alongside beings encountered within the world. When this unity is considered as a totality, it is understood as coming to its end, i.e., Dasein's death. Finally, that which gives Dasein to understand its transcendence as well as its finitude and "guilt" and, thus, calls it to achieve its own self is what Heidegger calls the *voice of conscience*. To achieve itself Dasein must let itself be called toward its genuine self, i.e., the process of finite transcendence.

The act in and through which Dasein achieves authenticity is called *resolve (Entschlossenheit)*.

Heidegger finally shows how care itself is founded in *time,* insofar as the basic components of care, namely ek-sistence, thrownness, and fallenness, inherently refer to the three ekstases of time: future, past, and present. By transcending beings toward Being, Dasein comes to its true self *(Zu-kunft,* future); but this self is always already as having been thrown forth (past), and concerning itself with beings, thus making them present and manifest (present). Interpreted from the perspective of temporality, resolve manifests itself as retrieve *(Wiederholung);* it lets the process of finite transcendence become manifest as *historical.* By fetching itself back time and again, Dasein lets its own self be in terms of its authentic past and that of its people; in addition Dasein also is as constantly coming toward its authentic self. It is in this complex process that Dasein hands over to itself its own heritage and thus "finds" its true self.--Let us now turn to the essays which make up this book.

First of all it should be stressed strongly that what is offered here is not some kind of brief summary of the most important sections of *Being and Time.* This anthology is not a commentary on *Being and Time;* rather it is a companion to this important work. The authors who have contributed to this volume point to issues with which the reader should be familiar if he or she is to understand Heidegger's text correctly. In other words, the essays contained in this anthology are not meant to be a substitute for Heidegger's text; rather they are all intended to lead the reader to a proper understanding of Heidegger's basic concern and to assist them in coming to a meaningful encounter with this great work.

The first essay, "Signification and Radical Subjectivity in Heidegger's *Habilitationsschrift,"* by Roderick M. Stewart takes us back to 1916, the year in which Heidegger officially concluded his education. Shortly thereafter he devoted himself to his research and teaching, first in the capacity of a *Privatdozent,* later as *extraordinarius* (1923), and finally as *ordinarius* (1928). Stewart's essay shows that in 1916 Heidegger was trying to break through the conceptual framework in which he had been educated at the University of Freiburg, namely neo-scholasticism and neo-Kantianism. In his *Habilitationsschrift* Heidegger shows that he has a thorough understanding of the dominant philosophical trends of his time, that he is dissatisfied with the intellectual world to which they had led him, that he is in the process of overcoming

this complex framework, and that in this latter effort his familiarity with Husserl's phenomenology plays an important part. Of
this and his other early works Heidegger later wrote that they
were help-less efforts which show that he still knew nothing of
what would concern him in his later thinking. Yet in his view,
they nevertheless also show a beginning path, even though that
path was at that time still largely hidden to him.

    Between 1916 and 1919 some very important events must have
taken place in Heidegger's life. Not much is known about these
formative years. It is a fact that in 1919 ideas begin to appear in
Heidegger's courses which have almost no relationship to his
earlier thoughts. The second essay contained in this anthology,
"Heidegger's Early Lecture Courses," by Theodore J. Kisiel
focuses on the most important ideas that began to surface in Heidegger's lectures which he taught first in Freiburg and later at
the University of Marburg. In this fascinating story Kisiel shows
how the great ideas that finally would constitute the core of
Being and Time, begin to come to the fore in the courses mentioned. Some of these courses have already been published;
others have been promised and will appear in the relatively near
future. Drawing on the early lecture courses insofar as we have
access to them today, Kisiel provides us with an overview of Heidegger's development during these years and focuses particularly
on two basic notions: Heidegger's first conception of a hermeneutics of facticity and his tentative response to the question of
whether the return to origins in philosophy should be scientific
or "historical."

    In his essay, "Heidegger's 'Introduction to the Phenomenol-
ogy of Religion', 1920-1921," Thomas Sheehan explains the influence which Heidegger's vast knowledge of Christianity and Christian theology had on certain basic themes developed in *Being and
Time.* In Sheehan's opinion *Being and Time* is "a creative mix
which at one and the same time appropriated and expropriated the
whole philosophical tradition." Some of the sources of Heidegger's
inspiration are well known by now; first there are the major
sources of inspiration: Plato, Aristotle, Kant, Hegel, Dilthey,
Husserl; then there are the secondary sources, such as Jaspers,
Kierkegaard, Augustine, Bergson, Scheler, Nietzsche, and
others. Until recently another major influence on *Being and Time*
remained concealed in the background, namely Heidegger's reading of St. Paul and his knowledge of early Christianity.

    Sheehan focuses only on one element of this broad and com-
plex topic, namely on the lecture course *Einleitung in die Phä*

nomenologie der Religion which Heidegger delivered in Freiburg in the winter semester of 1920-1921. To place the discussion of the themes developed in this course in their proper historical perspective Sheehan first provides the reader with some information on Heidegger's religious orientation in the years between 1914 and 1920. In the second section of his article Sheehan discusses two basic questions which Heidegger raises in his lectures on the phenomenology of religion: 1) what does Heidegger understand by "factical life-experience"? and 2) how can Christianity, interpreted as factical life-experience, be understood as primordial temporality? He also explains in what sense the themes developed here prepared some basic insights later systematically developed in *Being and Time,* particularly his new doctrine of temporality *(Zeitlichkeit).* Sheehan concludes his essay with a brief discussion of the relationship between Husserl and Heidegger between 1917 and 1920, a relationship already touched upon in the first section of the article.

Jacques Taminiaux's essay, "Heidegger and Husserl's *Logical Investigations.* In Remembrance of Heidegger's Last Seminar (Zähringen, 1973)," discusses Heidegger's relationship to Husserl and Husserl's phenomenology. The author explains that this relationship is very complex and even ambiguous. On the one hand, Heidegger recognizes his great debt to Husserl and his work; on the other, Heidegger also demands the right to develop these ideas in such a way that they become relevant to what he saw to be the basic issue of his entire philosophical efforts, namely the question concerning the meaning and truth of Being. Taminiaux also shows why Heidegger was so fond of Husserl's *Logical Investigations,* and why he was rather critical of Husserl's later transcendental phenomenology. Finally the essay describes in what sense Heidegger thought that in the sixth chapter of the Sixth Logical Investigation ideas are found about categorial intuition which would help him find an answer to the question raised in Brentano's book *On the Several Senses of Being in Aristotle.*

"The Origins of Heidegger's Thought" by John Sallis was written at the occasion of Heidegger's death in 1976. Sallis takes the occasion of Heidegger's death as a starting point for reflections on the origins of Heidegger's thought. It appears that the question concerning the origins of Heidegger's philosophy is to be answered at three progressively more fundamental levels. These levels appear to correspond to three different concepts of origin. In the first part of his article Sallis takes "origin" in the sense of *historical* origin and thus inquires about those thinkers whose works were decisive for Heidegger's development. In the second section of the essay origin is taken in the sense of original and *basic issue*. According to Sallis this basic issue is disclosedness and, thus, the author shows in detail how this issue serves as origin. Finally, the term "origin" is taken in its most radical sense as *that which grants* philosophical thought its true content. In the third and final part of his essay Sallis explains in what sense truth is the self-withdrawing origin of thought. There the author also explains that at this deepest level death regains a signifying power for philosophy which, all differences notwithstanding, could match that which it had among the Greeks.

The next essay is written by John Caputo and is entitled "Husserl, Heidegger and the Question of a 'Hermeneutical' Phenomenology." Caputo begins by describing a paradox in Heidegger's conception of the method of philosophy: on the one hand, he claims that phenomenology (which is this method) is descriptive; on the other hand he states that it is hermeneutical and thus interpretative. But how can one and the same method be at the same time both descriptive and interpretative? In disentangling this paradox Caputo is led to a very thoughtful reflection on the relationship between Husserl's pure and transcendental phenomenology and Heidegger's hermeneutic phenomenology. The author shows that the common description of the difference between Husserl's and Heidegger's phenomenology according to which Husserl searches for a presuppositionless form of thinking, whereas Heidegger does not wish to get rid of one's presuppositions, but rather to discover the right presuppositions which indeed illuminate the "things themselves," is much too facile a way to differentiate Husserl and Heidegger. Caputo shows with great care that in Husserl's phenomenology there is an essential hermeneutical component to which Husserl has devoted numerous and typically painstaking analyses. Caputo finally explains that the true difference between Husserl and Heidegger is to be found in the fact that Husserl's phenomenology contains *ontological* presuppositions which he "refuses" to examine critically, whereas Heidegger precisely focuses on such a critical analysis of these *ontological* presuppositions.

In his essay, "Heidegger and the Destruction of Ontology," Samuel IJsseling begins by giving a clear characterization of what Heidegger understands by phenomenology; for Heidegger phenomenology implies three basic components: phenomenological reduction, phenomenological construction, and phenomenological destruction. After explaining what is meant by these technical

expressions the author then focuses on the meaning of the phenomenological destruction. According to IJsseling, Heidegger has
expressed himself concerning the meaning of the destruction on
several occasions; yet the project nevertheless remains somewhat
ambiguous even though it plays a very important part in Heidegger's thinking as a whole. IJsseling first explains the function of
the destruction in the context of a fundamental ontology of the
kind developed in *Being and Time.* He then explains in what
sense destruction is to be related to reduction and construction
and what role it has to play in the retrieval of the Being-
question. Finally the question is asked of what the function of
the destruction is in Heidegger's later work, where he is concerned with an effort to overcome metaphysics by means of a step
back "out of the already thought into an unthought whence what
has been thought receives room to issue and abide in its
essence." (ID, 44/48) In these reflections IJsseling has made use
particularly of Heidegger's book, *The Basic Problems of Phenomenology,* which contains the text of a lecture course presented in
Marburg immediately after the publication of *Being and Time.*

In "Being-True as the Fundamental and Basic Determination
of Being," I have made an effort to describe Heidegger's interpretation of Aristotle's conception of truth which plays an important part in several lecture courses delivered between 1925 and
1930. In so doing I have tried to make a contribution to the
problem of how Heidegger in 1930 conceived of the relationship
between the question concerning the meaning of Being and the
question of truth, taken as *a-letheia.* It seems to me that a discussion of these issues will prepare the reader for the conception
of truth which Heidegger unfolded for the first time publicly in
section 44 of *Being and Time.*

Richardson's essay, "Heidegger and the Quest of Freedom,"
focuses on Heidegger's conception of man's freedom, an issue that
is very closely related to Heidegger's concern with truth.
Richardson's essay was originally written for *Theological Studies.*
Both in the introductory section and the conclusion the author
refers to the relevance of these reflections for contemporary moral
theology in general, and the debate about moral issues in the
Catholic Church in particular. These references may not be pertinent for many readers of this anthology. Yet it seems to me that
they clearly show the actuality of some of the *philosophical* issues
raised by Heidegger. At any rate, Richardson explains in detail
the intimate relationship which, for Heidegger, exists between
Dasein's transcendence, freedom, and truth.

For Heidegger time is the possible horizon for any comprehension whatsoever of Being. He stresses time and again that in
reflections on the meaning or truth of Being, time plays the
essential part; this is the case not only for his early works, it is
also true for his later work. It is thus of the greatest importance
for the reader of *Being and Time* to understand properly what
Heidegger understands by time, Dasein's temporality, and the
Temporality of Being, and how his new conception of temporality
(Zeitlichkeit) relates both to the everyday conception of time
which has been thematized in classical metaphysics since Aristotle,
and to the temporality *(Temporalität)* of Being itself. This is the
reason why I have decided to include two essays on Heidegger's
concern with time. In her essay, "The Concept of Time in Heidegger's Early Works," Marion Heinz focuses first on the starting
point, the method, and the structure of Heidegger's early investigations about time. She then discusses in detail Heidegger's view
on the temporality which constitutes the Being of Dasein. Finally,
she relates this conception of temporality to Being's own Temporality and explains how these new reflections on time are to be
related to our "ordinary" conception of time. Graeme Nicholson
begins his essay, *"Ekstatic Temporality in* Sein und Zeit," with a
brief explanation of the intimate relation which exists between the
search for the meaning or truth of Being, on the one hand, and
the question concerning the Being of time, on the other. Nicholson then describes in detail in what sense for Heidegger the very
Being of man consists in care *(Sorge)*. It is the effort to give a
foundation to this conception of the Being of Dasein which led
Heidegger to his reflections on temporality and time; this is
explained in detail in the second major part of Nicholson's essay.
The author next focuses on the temporal predicates with the help
of which Heidegger characterizes the various modalities which the
three ekstases of time assume depending on Dasein's mode of
care. The new names for each ekstasis of time are derived from
the analysis of care; yet the overarching form of ekstatic unity
into which these new predicates are brought, does not stem from
the analysis of care, but is rather imposed by the analysis of
temporality. Nicholson then describes the three ekstases of time
in detail and explains their typical unity. He also discusses the
variations in the temporality of Dasein which are the result of
Dasein's attitude and orientation, of Dasein's choices regarding
the way he will live in the world. Even though each form of care
thrusts toward one ekstasis in particular, it nevertheless also
implicates Dasein in the other two ekstases; but this phenomenon

is independent of Dasein's choice. The essay ends with a brief summary of the most important issues discussed.

The anthology concludes with a well-known essay by Otto Pöggeler, "Metaphysics and Topology of Being." The reason for including this essay in this ontology and for placing it at the end of the book is to be found in the fact that it gives the reader a clear idea about the relationship between Heidegger's earlier and later works. The author describes Heidegger's basic concern, discusses his development which was to lead to *Being and Time* and explains the reasons which led Heidegger eventually beyond the position reached in *Being and Time*. Pöggeler admits that Heidegger's basic concern and development can be portrayed in more than one way, and that one can use different titles for such an interpretation. He himself chose to depict this concern and development in terms of a movement from a metaphysics of beings to a topology of Being.

As far as the format of this anthology is concerned I have made an effort to respect the choices concerning the technical vocabulary as well as the style of the authors who have contributed to this book as much as possible. This is also the reason why I have not attempted to establish a complete uniformity as far as the translation of technical terms and expressions is concerned. Some authors prefer to translate the German term *"Sein"* by *Being*, whereas others prefer to avoid the use of capitals where they are not absolutely necessary. *Seiendes* is translated by some authors by *being,* whereas others prefer to use the term *"entity."* There still are authors who prefer to stay as closely as possible to the terminology introduced in the English translation of *Sein und Zeit* by Macquarrie and Robinson; others prefer to develop their own technical vocabulary. With respect to notes and references I have respected the different practices of the contributing authors wherever possible. It seems to me that even though this is a flaw if looked at from an aesthetic point of view, it should nevertheless not cause any serious problem for the attentive reader.

I have added a selective bibliography to this volume in which I have listed all the works by Heidegger quoted or used in the essays contained in this book as well as all the secondary sources mentioned by the different authors. I have added a few works of a more general nature which may be helpful to those who are not yet very familiar with Heidegger's thought, as well as a list of other anthologies on Heidegger's philosophy.

I wish to thank cordially the authors who have contributed an essay to this volume, the publishers and editors who have generously granted permission to use material which already had appeared elsewhere, as well as Ms. Jeanette Walther who so elegantly composed this book on her IBM Displaywriter.

Joseph J. Kockelmans
The Pennsylvania State University

1. RODERICK M. STEWART

SIGNIFICATION AND RADICAL SUBJECTIVITY IN HEIDEGGER'S HABILITATIONSSCHRIFT*

"The doctrine of signification [*Bedeutungslehre*] is rooted in the ontology of Dasein. Whether it prospers or decays depends on the fate of this ontology."[1]

I

Heidegger's *Denkweg* or "path of thinking" about the *Seinsfrage*, especially to the extent that it is a question about the human capacity for "discourse" (logos) and "language", takes its beginnings almost two decades before he wrote his first major work, *Sein und Zeit*.[2] The first published[3] signs of these concerns are some early articles and reviews, his 1914 doctoral dissertation and, finally, his "Habilitation" dissertation and "trial lecture" for the University of Freiburg in the period 1915-1916.[4] I shall narrow my focus in this essay to some themes and tensions that emerge in this *Habilitationsschrift* as Heidegger attempts to assimilate a medieval signification theory to then-available neo-Kantian logical theory and budding Husserlian intentionality theory. This discussion should enable us to see a clear anticipation of Heidegger's account of Dasein or radical subjectivity in SZ.

 The key to making this interpretation lies in pitting the "underlabourer" discussion of the main body of the *Habilitationsschrift* against the somewhat rebellious or restless tone of its final chapter or *Schlußkapitel*. The latter was written by Heidegger as

*From Roderick M. Stewart, "Signification and Radical Subjectivity in Heidegger's *Habilitationsschrift*," *Man and World*, 12 (1979), pp. 360-377. Reprinted by permission of the author and Martinus Nijhoff Publishers, B.V.

1

a "supplement" *after* all the formal pressures for entrance to the Philosophy Faculty of the University of Freiburg. But before we examine the signs of rebellion and unrest in this *Schlußkapitel*, let us set the stage by way of a brief review of the main body of the text.

II

The *Habilitationsschrift* is entitled, *Die Kategorien- und Bedeu-tungslehre des Duns Scotus*.[5] Though in fact the texts discussed by Heidegger are not by Scotus, but Thomas of Erfurt (as Martin Grabmann has shown), we may disregard this scholarly point in favor of the philosophical spirit of Heidegger's interpretation.[6]

Part I of the *Habilitationsschrift* is entitled, "The Theory of Categories: Systematic Foundations for Understanding the Theory of Signification," and its purpose is to show "Scotus'" awareness of the irreducibly different categories or kinds of objects (*entia*) above and beyond those Aristotelian categories applicable only to spatio-temporal, physical substances. In Husserlian language the goal is "regional ontology." Heidegger reviews "Scotus'" discussions of two of the most general categories predicable of any object (*ens*) whatsoever; the "transcendental" One (*das Unum*) in Chapter I (FS, 149-207), and the True (*das Verum*) in Chapter 2 (FS, 207-231).

Chapter 1 rearticulates "Scotus'" distinction between primitive and derivative senses of *unum* (namely, "one as opposed to another" *versus* "the number one") in light of work by the senior neo-Kantian philosopher and logician at Freiburg, Heinrich Rickert.[7] A clear distinction is "shown" (*aufgewiesen*), though not "proven" (*bewiesen*), between ideal reality (of logical and mathematical sorts) and empirical reality.

Chapter 2 interprets the medieval maxim that every object as such is a true object into the Neo-Kantian axiom that all objects must be objects of knowledge, and all knowledge must be of objects. Moreover, in this Kantian tradition such knowledge is only possible in *true* (predicative) judgments. The object *as known (ens logicum)* becomes, then, a judgmental-content, an abstract (intensional) entity which must be distinguished ontologically from any particular mental act of judging.[8] A further categorial or ontological distinction is drawn--this time between psychologically real events and ideal "logical" entities.[9]

Yet, what Chapters 1 and 2 of Part I put asunder, Chapter

3 and Part II of the *Habilitationsschrift* seek to join together. The contents or senses of judgments, whether about natural objects or ideal mathematical and propositional entities, are all *expressible*. They find their way into words and are communicated and "signified" by them. Hence, a theory of signification (*Bedeutungslehre*) comes to the fore. And for Heidegger's interpretation of "Scotus," this signals a further assimilation of this medieval text to the concepts of Husserlian phenomenology, particularly as found in Husserl's revived idea of a universal grammar in the *Logische Untersuchungen*,[10] and in his theory of noetic acts, noematic-senses and intentional objects, as found in the *Ideen* of 1913. But before we summarize the results of Heidegger's interpretation of this signification theory, a few words need to be said about this Husserlian terminology.

The basic unit of Husserl's general theory of intentionality is an "intentional act or lived-experience (*Erlebnis*)." This is not essentially an act or activity in a literal sense, but rather an abstractable event from someone's "mental life" (however this is to be explained metaphysically). Let us say these acts are "designated by" sentences like the following:[11]

1) Smith sees his Cape-Cod house.
2) Jones imagines a unicorn.
3) Brown thinks of the number 9; of the fourth root of 6561.
4) Williams judges (out loud or to himself) or asserts that his Cape-Cod house is grey.

Ignoring the full complexities and varying terminology in LU and the *Ideen*, we may distinguish the following moments of each of these acts or events taken in its entirety:

i) the noetic act (type) or "act-quality": e.g., seeing, imagining, thinking, judging.

ii) the "intentional object(ive)" (*Gegenständlichkeit* in a wide sense): e.g. the physically existing Cape-Cod House, the ideally existing number, the state-of-affairs that the Cape-Cod House is grey.

Note: In Rickert's theory of judgments, for cases like (4) above, this is roughtly equivalent to the "transcendent sense" of each judgment. These are the genuine "objects of knowledge" for Rickert.

iii) the noematic-sense or "act-matter": e.g., the perspective under which an object is perceived or imagined; the description by which we think of an ideal object. If Brown thinks of the number 9, then of the fourth root of 6561, each act is *of* the same object, but is distinguished by its noematic-sense.

Note: In the case of judgment, this is referred to as a judgmental content or sense; what Rickert called the "immanent sense" of the judgment.

We shall employ these distinctions when we summarize below Heidegger's interpretation of "Scotus'" signification theory. But first we need to get "Scotus'" own story straight.

A theory of signification, as a *Grammatica speculativa* or universal grammar, for "Scotus" specifies (for any possible language) the most elementary grammatical categories of signifying objects and their properties by terms (nouns, verbs, adjectives, etc.), and the *a priori* rules according to which these categories or their instances may be combined[12] into more complex meaningful or signifying units, or semantically uninterpreted, but syntactically well-formed formulae.[13]

The following story takes place for any given simple term to be able to signify an object and its properties. According to "Scotus," there must first exist in itself a thing with a property or mode of being (*modus essendi rei*); what we eventually say is determined ultimately by how things are. Heidegger will call this the principle of the "material determinateness" of every form. (FS, 253) Then, the "intellect" (*intellectus*) apprehends a property of a thing (*res, proprietates = species intelligibilis*) and on this basis assigns (*tribuit*) to a certain sound (*vox*) the active property of signifying the thing and its property (*modus significandi activus*). The *vox* becomes a *pars orationis* if it signifies in the specific form of noun, verb, etc. (FS, 250, n. 4) The thing takes on the passive property (*modus significandi passivus*) of being signified in just the way specified through the *vox* by *intellectus*. (FS, 251, n. 7) But, in order for the intellect to set up such signification by a *vox*, it must first apprehend or understand to some degree the thing and its property; consequently, *intellectus* acquires the active property of knowing the object (*modus intelligendi activus*) and the thing acquires the passive property of being known to whatever degree it is (*modus intelligendi passivus*). (FS, 258, nn. 18,20). Finally, we should note "Scotus'" claim that the two passive *modi* are "materially identical" but "formally different". Each is that mode or property of the *same* "matter" (namely, the *res*) that is, formally speaking, either signified or known by the intellect. "Scotus'" theory gets translated into the intentionality-idiom, but not without some interpretative tailoring. "Intellect" comes to be regarded as "consciousness" (*Bewußtsein*). The *vox* endowed with meanings is taken as the Husserlian technical term, "expression" (*Ausdruck*), the physically written or spoken word(s) endowed with meaning by significative acts of consciousness. However, the *modus significandi activus* is no longer a property of the *vox*; it is ambiguously treated by Heidegger as either the whole intentional act of signification *via* some *vox*,[14] or as the psychologically real (*reell*) counterparts to the abstract qualitative and material moments of the act of signification.[15] Moreover, according to Husserl's theory of expressions (LU I,§ 7, 33 (277)), the *vox* endowed with meaning signifies not the object itself with the property, as "Scotus" suggested, but rather the speaker's "thoughts".

On the other hand, Heidegger's rendering of the passive mode of signifying is unproblematic. For "Scotus," it was a property of the *res*, that it underwent being signified by the *vox*. As with all the passive modes, Heidegger interprets the *modus significandi passivus* as the *intentional content* of the act of signification (FS, 262), the noematic-sense of the act (252), the "objective correlate of the act" (251), the act-matter (253). Or, alternatively, it is the transcendent thing "to the extent that it is related to expressions, i.e., to the extent that it has entered into significations." (FS, 261) In other words, the noematic-sense or content of an act of signification is the intentional object just as signified under a certain description.

"Scotus'" *modus essendi rei* is, according to Heidegger, (ii) above: that objectivity referred to by the noematic-sense (FS, 252). Moreover it can be, as for Husserl, "whatever is objective in general" (FS, 256) and, consequently, comprehends natural, imaginary and non-sensory object(ive)s.[16] Finally, Heidegger makes an adequate accommodation of "Scotus'" reference to the property of the *res* as a *species intelligibilis*, a knowable (specific) form. The *modus essendi rei* may be rendered into Husserlian language as "what is experienceable in general" (*das Erlebbare überhaupt*). (FS, 260)

Heidegger also provides a direct paraphrase of the *modus intelligendi activus*. It is "the manner and fashion in which I grasp something objectively and thereby know it." (FS, 258) In other words, it is a property or mode (an act-qualitative moment)

of consciousness or intellect; that specific quality which makes
the concrete act an act of understanding, knowing rather than,
say, an act of signifying. Moreover, the *modus intelligendi
passivus* is also interpreted consistently with the "Scotus"-text:
"The passive mode [of knowing] is nothing other than the *modus
essendi* insofar as it has been objectified in consciousness." (FS,
259) In other words, the passive mode of being known is that
property of the thing that is known by the intellect.

But how does Heidegger accommodate the "material identity"
and "formal difference" of the various *modi*? Heidegger takes this
to mean that the noematic-senses of *three* different acts are identical, while the overall description of each of these acts (the
whole noema of each act taken in reflection) differs. We may
understand this as follows. Suppose it is a fact that 1) S (*res*) is
P, i.e. the mode of being of some S is P: namely, this Cape-Cod
House is grey. Suppose further that 2) Jones (in an act of knowing) judges (without saying so) that this Cape-Cod House is
grey. Finally, suppose that 3) Jones (in a significative or
"expressive" act) asserts or says on the basis of 1) and 2) that
this Cape-Cod House is grey. The noematic-sense or act-matter of
the act of silent knowing and the significative act is the same
'that this Cape-Cod House is grey'; hence, both acts are (act-)
materially identical and, in this, according to Heidegger, the
modus significandi passivus and the *modus intelligendi passivus*
are materially identical.

But how can Heidegger account for the fact that these two
passive modes are identical to the *modus essendi rei simpliciter*,
as "Scotus" claimed? As it turns out, Heidegger's commitment to
the intentionality-idiom and to the Kantian principle that all
objects are objects of knowledge requires him to "improve" upon
"Scotus" and suggest both a passive and an active *modus essendi
rei*![17] On "Scotus'" terms this would make no sense. For Heidegger, this move amounts to distinguishing a predicative or judgmental act of knowing an object under some specific description
(See 2) above) from that intentional act in which the "absolute
objective reality...stands under a specific *ratio*...of existence,"
or that [act] in which the immediate givenness actually comes into
consciousness." In Kantian language, this new act is that of
being immediately (intuitively) aware of the *res* and its property
or mode of being. In keeping with our previous distinctions, we
may then suppose that 4) Jones is immediately (pre-predicatively)
aware of this grey, Cape-Cod House. Here 'being immediately
aware of' designates the act-qualitative moment, the *modus*

essendi activus, and 'this grey, Cape-Cod House' the act-material moment, the *modus essendi passivus*. If we assimilate the explicit predicative relation in 2) and 3) to the "inherence" relation implicit in 4), the act-matter of this act is virtually identical to those of 2) and 3). Each act, however, is "formally different" in that it "achieves" (*leistet*) different things as is indicated by the varying act-qualities.

One final note: For "Scotus" the *modi significandi* presupposed the *modi intelligendi*, which in turn depended on the *modus essendi rei*; how and what terms signified depended on what the intellect could know of how things in fact are. Heidegger translates this "entanglement" (*Verschlungenheit*) or "meshing-together" (*Ineinandergreifen*) of the various *modi* into Husserl's language of one kind of intentional act "founding" or being "founded on" another kind, with the *modus essendi* as the "final fundament" (in the order of being) of all these. (FS, 259)[18]

We are finally in a position, I think, to examine the change in tone of the "supplemental" final chapter of this "Scotus"-writing. The previous discussion should suffice to indicate how well-entrenched Heidegger's own discussions in the main body of text were in the related conceptual schemes of neo-Kantian logic and Husserlian phenomenology.

III

The "Scotus"-writing is dedicated to Heinrich Rickert "in most grateful veneration." In fact, two years before in the Preface to his doctoral dissertation on psychologism and judgment theory Heidegger wrote:

> No less shall I remain indebted with grateful mind to Privy Councilor Professor Rickert. Him I thank for my seeing and understanding the problems of modern logic. (FS, 3)

Yet, two years later in the Preface to the *Habilitationsschrift* on "Scotus", a restless tone appears:

> This dedication is the expression of indebted gratitude; but at the same time, if one is to safeguard one's own 'standpoint' in a fully free fashion, its intention is to make known the conviction that the problem-conscious, world-view character of [neo-Kantian] *Wertphilosophie* has been called to a crucial

forward motion and deepening of the way it handles philosophical problems. (FS, 133)

And, finally, in the opening paragraphs to the final chapter:

...this is now the appropriate place to express the mental
unrest, until now suppressed, that the philosopher experiences in his world of problems each time his studies take on
an historical shape. (FS, 342)

This "mental unrest" shows up, as we shall see, with a new
notion of subjectivity as *lebendiger Geist*.
    This final chapter is entitled, "The Problem of Categories,"
and "attempts to provide preliminary stipulations concerning the
structure of the problem of categories and a possible way for
solving it." (FS, 354) The "possible way for solving" the problem
of a theory of categories that are employed in thinking of objects
in general (and for signification- and judgment-theory) is to take
seriously a certain theory of subjectivity or mind. In fact, it is
to be urged, such a "philosophy of mind" ought to be the cor-
nerstone of all metaphysics and philosophy.
    In this vein, Heidegger proposes three fundamental tasks or
requirements for a "promising solution to the problem of categories" raised by his study of the (first part of the) "Scotus"-text.
The first requirement or task is to provide a "characterizing demarcation of the various domains of objects into sectors [*Bezirke*]
which are categorially irreducible to one another." (FS, 342) This
first requirement has been met within the framework of the
"Scotus"-text, but to the exclusion of "deeper-reaching metaphysical sets of problems." The "deeper-reaching" problems themselves present a second task: "the situation of the problem of
categories within the problem of judgment and the subject." (FS,
343) Yet, as we shall see with the notion *lebendiger Geist*, the
"problem of the subject" is more radical than that of a cognizing,
judging subject or even a "pure, intentional consciousness." A
third requirement, then, is that subjectivity as *lebendiger Geist*
"is only to be comprehended when the whole abundance of its
performances [*Leistungen*], i.e., *its history*, is brought into
relief." (FS, 350)
    The first task is clearly addressed to the neo-Kantian and
Husserlian reader. Category theory is not just a theory for
thinking, but for thinking about objects. Since some categories
(what Rickert's student, Emil Lask, called "reflexive" ones) are

alleged to apply to any object, however intuited, and others only apply to empirically intuitable ones, we are led to a natural division between domains of objects: empirical and non-empirical ones. And we may give (partial) essential descriptions of each of these: members of the one are spatio-temporally extended, the others are not. Moreover, we may find that each division has sub-divisions: on the one hand, perhaps "animate" and "inanimate", etc.; on the other, logical and mathematical, etc., in each case supplying appropriate *differentiae*. Each division gives us a "sector" of possible intentional objects under general (eidetic) descriptions; each successive sub-division gives us a more *specific* perspective (noematic-sense) on the object. In principle, we get an inventory of the possible objects for consciousness. Typically, of course, such "regional ontology" or category theory concerns itself with the most general descriptions (the *transcendentiae* of any *ens* for "Scotus") under which any objective may be thought.

Heidegger discusses the second task for a "promising solution to the problem of categories" as follows:

We may only understand these [previously excluded metaphysical questions] as being ultimately decisive for the problem of categories if we recognize a second fundamental task for any theory of categories: situating the problem of categories within the problem of judgment and the subject....Admittedly, the presentation of Duns Scotus' theory of judgment had another tendency: it was supposed to characterize the domain of what is logical, and in this the essential relation of the judgment to the category remained completely in the dark. On the other hand, the theory of signification did permit some access to subjectivity (by which we do not mean individuality but the *subject in itself*). The task of Duns Scotus, the analysis of a certain layer of acts, the *modi significandi*, forced him to enter the sphere of *acts in general* and to establish certain fundamental things about the individual layers of acts (*modus significandi, intelligendi, essendi*) and their relation to one another.

Precisely the existence of a signification theory during medieval scholasticism reveals a fine disposition for accurately giving ear to the immediate life of subjectivity and the nexus of meanings immanent to it, without having acquired a sharp concept of the subject. (FS, 343)

The second task is actually two-fold: on the one hand, to establish the connection of category theory to judgment theory; on the other, to determine its connection to an appropriate theory of subjectivity. The first part of the task was never clearly articulated in his discussion of "Scotus'" *transcendentiae* for any *ens* or of his doctrine of the various *modi*. The connection is this: the structure of any judgment is always predicative, with a logical subject and logical predicate. The "categories" form an important subset of these predicates, namely, the most general ones, though we may distinguish kinds of categories appropriate to all or only some modes of intuiting particulars. (FS, 345-346)

However, the categories must not be thought of as mere forms or "functions of thinking," but as forms necessarily bound up with a "matter," namely, objects. Heidegger invokes here the neo-Kantian principle that "emphasizes from the outset that all thinking and knowing is always thinking and knowing *an object*." (FS, 345) This "material determinateness" even applies to Lask's so-called "reflexive" categories.[19] Heidegger asks, citing Külpe and Rickert, whether this means that the only objects we even know and to which we may *validly* (with *Geltung*) apply our categories are the "immanent" appearances of our subjective experiences of the world and not the "transeunt" things-in-themselves? (FS, 346) His answer serves to introduce the second part of the second task--locating the category problem in the problem of the nature of subjectivity:

It is indisputable "that all transeunt *Geltung* stands and falls with the recognizing of objects" (O. Külpe, *Zur Kategorienlehre*, p. 52); only, the problem is precisely to know what sort of objectivity this can be, once one considers the fact that objectivity makes sense only for a judging subject; without this subject, we could never succeed in bringing out the full sense of what we designate by the term 'Geltung.' (FS, 346-347)

Referring to neo-Kantian discussions about the status of ideal, "geltende" objects, Heidegger continues:

We need not decide here whether it [*Geltung*] refers to a peculiar 'Being' or an 'Ought', or to neither of these, but is only to be comprehended by means of deeper-lying groups of problems which are embraced in the concept of living Spirit and unquestionably are closely connected with the problem of Value. (FS, 347)

The problem in this passage is that Heidegger begins by talking about the *Geltung* of the categories for "immanent" or for "trans-eunt" objects--whether the categories hold-true-of (*Gelten von*) only "immanent" appearances or also of "transeunt" things in themselves. His answer seems to be that the only "transeunt" objects that can be made sense of are those that can in principle be known by the judging subject. The problem arises whether we must divide these knowable transcendent objects (or, in Rickert's language, "senses") into different modes of being or ontological types: empirical objects and non-empirical ones. However, at this point Heidegger uses 'Geltung' not to describe that feature of the categories when they "hold-true-of" various kinds of objects; rather, he uses it in a second way to describe the ontological type of non-empirical objects. Rickert had, before Heidegger, queried whether to call those ideal entities that constitute the proper *Gegenstand der Erkenntnis* entities with a peculiar form of "Being," or that belonged in the realm of the "transcendent Ought."[20] In any case, as far as Heidegger is concerned, whether Rickert's description is adequate depends rather mysteri-ously on solving some unknown "groups of problems which are embraced in the concept of the living Spirit and unquestionably are closely connected with the problem of Value." And it is with this that the second part of the second task is announced.

But what is the status of a theory of subjectivity in which the subject is to be understood as *lebendiger Geist*? Heidegger's remarks are very sketchy here; however, it is clear that this is first of all a full-blown metaphysical theory:

We cannot view logic and its problems in the true light at all if the context from which it is interpreted is not a translogi-cal one. *Philosophy* cannot dispense for long with its own optics, Metaphysics. (FS, 347-348)

If we contrapose the first sentence we get roughly, "if the con-text from which it [logic] is interpreted is a translogical one, [then] we can view logic and its problems in the true light." In other words, the "translogical context" or "light" by means of which we are to view "logic," as including signification-, judgment-, and *Wissenschafts*-theory, is best measured by the "optics" of philosophy, metaphysics.

But how does a study of *lebendiger Geist* provide such a metaphysical "optics" or translogical theory? Heidegger continues:

For a theory of truth, this means that we must give consciousness a final, metaphysical-teleological interpretation. What-is-of-value [*das Werthafte*] lives in consciousness in a profound and genuine fashion, insofar as it is an active, living deed that is itself full of meaning and actualizes meaning. None of this is understood in even the remotest way if it [this active, living deed of consciousness] is neutralized into the concept of a biological, blind matter of fact. (FS, 348)

What-is-of-value or holds-true is the set of "transcendent senses" or objects "intended" through all possible theoretical, practical and aesthetical judgmental contents (or, "immanent senses"). Though these ideal entities constitute for neo-Kantians the quasi-Platonic realm of *Geltung*, nevertheless they "live" in consciousness necessarily or "in a profound and genuine fashion" (i.e., if anything does at all) through its intentional (or, teleologically interpreted) acts. It would be a category mistake to treat these abstract "senses" as contingent by-products of the "blind" biological evolution of the human *lebendiger Geist*. A metaphysical-teleological view of consciousness treats them as "independently existing" ideal entitles or values. However they are instantiated in consciousness, as what-is-of-value, they make consciousness a "living Spirit," one that is "full of meaning and actualizes meaning" in all its dealings and "judgments".

Yet, all this talk of ideal entitles or values that are, in Rickert's language, the proper *Gegenstände der Erkenntnis* should not lead us to treat *lebendiger Geist* merely as a theoretical, judging subject. Heidegger cautions:

Within the wealth of productive or formative directions of the *lebendiger Geist*, the theoretical attitude is only *one*; for this reason we may call it a fundamental and disastrous error for philosophy as *Weltanschauung* if it satisfies itself with a spelling out of reality and does not aim for a *breakthrough* into true reality and real truth beyond a perpetually cursory synthesis and snatching up of the sum of what is knowable. Only by taking its orientation from the concept of *lebendiger Geist* and its "eternal affirmations" (Fr. Schlegel) will an epistemological logic be guarded from limiting itself exclusively to a study of structures, and only [by taking this orientation] will it make the logical sense, *even in its ontic meaning, into a problem*. Only then will a satisfactory answer be possible as to how "unreal" "transcendent" sense guarantees for us true reality and objectivity. (FS, 348)

The *lebendiger Geist* does not concern itself solely with Rickert's realm of the "transcendent Ought" or Lotze's "unreal" domain of *Geltung*. *Lebendiger Geist* has a "wealth of productive or forma-tive directions" which "break through" to the "ontic meaning" of these ideal structures, to the "true reality and real ruth" of immediately intuitable, or better, "encounterable," objects. It is interesting to note here that in subsequent paragraphs Heidegger commends Emil Lask for his work in judgment theory, where (with his notion of *Übergegensätzlichkeit*) he "breaks through" the study of "imitative," idealized propositional or judgmental sense-structures to a study of the "prototypes" of judgmental sense and truth given through sense-perception, and to how subjectivity "destroys" this prototypical realm through its judgmental synthe-ses.[21] Of course, this emphasis on discovering the "ontic mean-ing" of ideal "senses" and values in the "wealth of productive or formative directions" of *lebendiger Geist*, I would urge, has the definite ring of those passages in SZ where Heidegger distin-guishes Dasein's understanding and sense, as it is concernfully involved in its world, from the Cartesian, epistemological subject which is only accidently embodied and in commerce with a material world.

Now it is with this caution about our concept of subjectivity that Heidegger introduces the *third* fundamental task or require-ment for solving the problem of categories:

The epistemological subject does not explain the metaphysical-ly most significant sense of spirit, to say nothing of its full content. And only if it is situated within this sense will the category problem receive its proper deep dimension and enrichment. *The lebendiger Geist is as such an historical Geist in the widest sense of the word....Geist* is only to be comprehended when the whole abundance of its performances *[Leistungen]*, i.e., *its history* are set into relief;...History and its cultural, philosophical and teleological interpretation *must become* a sense-determining element for the problem of categories....Next to demarcating the domains of objects and the incorporation of the problem of judgment, this is the

third fundamental requirement for a promising solution to the
problem of the categories. (FS, 349-350)

Heidegger's "mental unrest" that we previously noted now gets
expressed in an explicit form. A metaphysical-teleological interpretation of a purely intentional consciousness is no longer adequate or "radical" enough for a full appreciation of the ontological
issues in medieval and neo-Kantian category theory. In the Introduction and in what follows the above passages (FS, 135-138;
350-353), Heidegger indicates that any purely formal approach to
the problem of categories is doomed to failure, since such formal
approaches do not take into account the culture and human
experience of the period in which the fundamental categories are
discussed. In fact, an appropriate understanding of medieval
category theory also requires a study of medieval mysticism
(e.g., Eckhart) and theology (e.g., a full appreciation of the
human and religious or onto-theological commitments of a concept
like "analogy"), and vice versa.[22] Moreover, a complete account
of category theory must comprehend the entire range or cultural
"history" of the unraveling of the living (human?) Spirit.

    But, as much as this distinction between a merely "formal"
and "historical" method of understanding sounds like Rickert's
distinction between "generalizing" explanation in the natural
sciences and "individualizing" understanding in the cultural
sciences (Cf. further the "trial lecture" on the concept of time in
the historical sciences: FS, 355ff.),[23] it is probably better taken
in the spirit of Wilhelm Dilthey's distinction between the natural
sciences and the human sciences (Geisteswissenschaften): this is
roughly a difference between Rickert's attempt to find more
objective criteria for understanding historical reality in a universal Kulturwissenschaft, and Dilthey's "historicist-relativist",
and hence more "subjective" approach to understanding history
and culture. At this time Heidegger had been very impressed with
the historicist and hermeneutical writings of Dilthey, as introduced to him by Carl Braig, a professor of theology at Freiburg
and the last representative of the Tübingen school of speculative
Catholic theology in its dispute with Hegel and Schelling.[24] Presumably, he was familiar with Dilthey's objection to neo-Kantian
philosophy of history and science that it had not adequately distinguished the "personal" empirical character of the historical
sciences from the concept of empirical verification in the natural
sciences.

    The influence of this Tübingen school would also explain

Heidegger's final remark that a philosophy of *lebendiger Geist* must "face the major task of a fundamental disagreement or show-down with...Hegel" (FS, 353), though the details of this *Ausein-andersetzung* are not provided by Heidegger. We may speculate as follows: if the account of Dasein's "historicity" in SZ in any way captures Heidegger's thinking at this time, then perhaps his showdown with Hegel centers around the proposal that there can be any single, universal essence or "true" conceptual description of human existence in a world as, say, *animal rationalis*—one that is slowly actualizing itself through the course of history. If this speculation is on track, then it lends further weight to the inter-pretative claim that a philosophy of world-historical *lebendiger Geist* anticipates the radical notion of subjectivity-based meta-physics found in the account of Dasein's "historicity" in SZ.

To close, let us consider Heidegger's own "later" perspective on his earliest published works:

[A]t the time of the writing of [these] literally help-less early efforts I still knew nothing of what would later beset my thinking.

Nevertheless, they do show a beginning path at that time closed shut to me [*einen mir damals noch verschlossenen Wegbeginn*]: the question of Being in the form of the Problem of Categories, the question of language in the form of a theory of signification. That both these questions belonged together remained in the dark. This darkness was not even intimated by the unavoidable dependency of the method of treatment of these [two questions] on the prevailing standard for all onto-logic, the theory of judgment. (FS, IX)

In part, our goal in this essay has been to "disclose" this *ver-schlossenen Wegbeginn*. Of course, our concerns have only been with one of these early, in hindsight, "help-less" efforts. More-over, our focus in large part has been narrowed to the questions of "language" and "Being" as these arose for Heidegger in an attempt to assimilate in underlabourer fashion a medieval signifi-cation theory to neo-Kantian epistemological logic and to budding Husserlian intentionality theory. Yet, as this last section has shown, the signs of a future philosophical revolution are clearly present, especially in the final, "supplementary" chapter of this *Habilitationsschrift*. Just how "help-less" these early efforts are, in light of Heidegger's later *Denkweg*, goes well beyond the scope of this essay. Their importance, however, cannot be doubted if we take the later Heidegger at his word. He cites a line from Hölderlin's Rhine-hymns (4th verse): "Denn, wie du anfängst, wirst du bleiben."[25]

NOTES

1. Martin Heidegger, *Sein und Zeit*, 11th ed. (Tübingen: Niemeyer, 1967), 166, hereafter cited as SZ. Additional references will contain English pagination in parentheses from the Macquarrie and Robinson translation (New York: Harper & Row, 1962): in this case, p. 209.

2. This has been discussed with varying emphases in the literature. Cf. in English: Albert Borgmann, "Heidegger and Symbolic Logic," in *Heidegger and the Quest for Truth*, ed. Manfred S. Frings, (Chicago: Quadrangle Books, 1968), 139–162; John Caputo's two articles, "Language, Logic, and Time," *Research in Phenomenology*, 3 (1973), 147–155; *idem*, "Phenomenology, Mysticism and the 'Grammatica Speculativa': A Study of Heidegger's 'Habilitationsschrift'," *The Journal of the British Society for Phenomenology*, 5 (1974), 101–117; Thomas A. Fay, "Heidegger on Logic: A Genetic Study of His Thought on Logic," *Journal of the History of Philosophy*, 12 (1974), 77–94; Cyril Welch, "Review of *Frühe Schriften*. By Martin Heidegger," *Man & World*, 7 (1974), 87–91. In German: Otto Pöggeler, *Der Denkweg Martin Heideggers*, (Pfullingen: Neske Verlag, 1963), Ch. 3, 67ff.; also, Karl Lehmann, "Metaphysik, Transzendentalphiloso-phie und Phänomenologie in den ersten Schriften Martin Heideg-gers (1912-1916)," *Philosophisches Jahrbuch*, 71 (1964), 333–367; Edgar Morscher, "Von der Frage nach dem Sein von Sinn zur Frage nach dem Sinn von Sein--der Denkweg des frühen Heideg-ger," *Philosophisches Jahrbuch der Görres Gesellschaft*, 80. Jahrgang. Zweiter Halbband (1973), 379–385; Rudolph Gumppen-berg, "Die transzendentalphilosophische Urteils- und Bedeutungs-problematik in M. Heideggers 'Frühe Schriften'," *Akten des 4. Internationalen Kant-Kongresses*, Mainz, Teil II.2 (1974), 751–761.

3. This is not to overlook the influence of Brentano's dissertation on Aristotle which Heidegger received in 1907 in the final year of his gymnasial studies. For more on this, see Martin Heidegger, "Mein Weg in die Phänomenologie," in *Zur Sache des Denkens* (Tübingen: Niemeyer, 1969), 81–90; also David Farrell Krell, "On the Manifold Meaning of *Aletheia*: Brentano, Aristotle, Heidegger," *Research in Phenomenology*. 5 (1975), 77–94.

4. The doctoral dissertation, the *Habilitationsschrift* and the "trial lecture" are all reprinted in Martin Heidegger, *Frühe Schriften* (Frankfurt a.m.: Klostermann, 1972), hereafter cited as FS. The latter lecture has recently been translated into English by Harry S. Taylor and Hans W. Uffelman, "The Concept of Time in the Science of History," *The Journal of the British Society for Phenomenology*, 9 (1978), 3-10. Of Heidegger's various early articles, one should consult his series of reviews, "Neuere Forschungen über Logik," *Literarische Rundschau für das katholische Deutschland*, Oct.-Dec. 1912, 465ff., 517ff., 565ff. A discussion of these reviews can be found in R.M. Stewart, "The Problem of Logical Psychologism for Husserl and the Early Heidegger," *The Journal of the British Society for Phenomenology*. 10 (1979), 184-193.

5. I.e., *Duns Scotus' Theory of Categories and Signification* (Tübingen: J.C.B. Mohr, 1916); reprinted in FS, 131-354.

6. Cf. Martin Grabmann, "Thomas von Erfurt und die Sprachlogik des mittelalterlichen Aristotelismus," *Sitzungsberichte der Bayerischen Akademie der Wissenschaften*, Munich, 1943. I concur with the opinion of most of the commentators already mentioned, including Grabmann's own enthusiasm for Heidegger's *philosophical* work, that Heidegger's *interpretation* of this medieval work is what is most important. Indeed, Heidegger himself refers to "Scotus'" treatise on signification and its place in the history of medieval philosophy as follows: "A more thorough, *historical* characterization of the treatise as regards the systematic tasks of a theory of signification is reserved for a special investigation. In what follows, we are only concerned with the *theoretical* understanding of the theory laid down in it." (FS, 24)

7. Cf. Heinrich Rickert, "Das Eine, die Einheit und die Eins. Bemerkungen zur Logik des Zahlbegriffs," *Logos* 2 (1911), 26-78.

8. Cf. FS, 217-231. In passages we shall not be explicitly concerned with in this essay, Heidegger discovers arguments in "Scotus'" text against locating logical reality, both judgmental senses and their constituent meanings, what "Scotus" calls *entia rationis/ ens logicum*, in the soul (*in anima*). "Scotus" further describes such an item as an *ens diminutum*, the being of which cannot be the proper object of metaphysics (as the "real" science of nature); i.e., it cannot be investigated according to spatio-temporal and causal principles because it is *insensible*. (FS, 217-229) It must not be regarded as a *prima intentio* like real, "first order" objects (of the "natural attitude"), but as a *secunda*

intentio, a "second order" object (of the phenomenological/logical attitude). (FS, 221) Included in this first group for "Scotus" were (real) mental items, such as *actus intelligendi*, which "pertinet ad considerationem libri de anima." But the real, psychological-empirical account of these *acts* is presupposed by *Logica* ("logic"), the logical account of their ideal or propositional contents (FS, 228). Heidegger characterizes this domain of the *Logicus* as that of *Gelten*, citing Lotze, Rickert et al., and describes its "secunda intentio" as the logical content (*logischer Gehalt*), judgmental sense (*Urteilssinn*), and the noematic sense (*noematischer Sinn*).

9. This is, of course, the distinction Heidegger was at pains to draw in his doctoral dissertation, *Die Lehre vom Urteil im Psychologismus*; reprinted in FS, 1-29. For more on this topic, see Stewart, art. cit.

10. Edmund Husserl, *Logische Untersuchungen* (Tübingen: Niemeyer, 1968), Bd. II, Erster Teil, Investigations 1 through 4, esp. 4, 294-342; hereafter cited as LU. In the English translation by Findlay (New York: Humanities Press, 1970), Vol. 2, 493-529.

11. The reader should consult important recent work on Husserl's theory of intentionality and meaning, especially as it concerns Frege's sense-reference distinction and commitment to the existence of abstract, "intensional" entities: Dagfinn Føllesdal, "Husserl's Notion of Noema," *Journal of Philosophy*, 66 (1969), 680-687; Richard Acquila, "Husserl and Frege on Meaning," *Journal of the History of Philosophy*, 12 (1974), 377-383; Richard Holmes, "An Explication of Husserl's Theory of the Noema," *Research in Phenomenology*, 5 (1975), 143-153; J.N. Mohanty, "On Husserl's Theory of Meaning," *Southwest Journal of Philosophy*, 5 (1974), 229-243; Smith & McIntyre, "Intentionality via Intensions," *Journal of Philosophy*, 68 (1971), 541-561; *idem*, "Husserl's Equation of Meaning and Noema," *The Monist*, 59 (1975), 115-132.

12. "Scotus" in fact calls this syntax for actively signifying terms a *constructio* subject to four kinds of Aristotelian principles (FS, 265ff.): (1) the *Constructibilia* or simple signifying terms; on Heidegger's analysis, as we shall see, the simplest noematic-senses bestowed to the simplest words. (2) A formal principle of unification (*unio*) of these elements; (3) Two efficient principles, (a) *Principium efficiens constructionis intrinsecum*: the *modi significandi* themselves. The terms (*voces*) may be combined with one another in dependent or independent ways in virtue of the way each term signifies the object (e.g., the difference between

nouns and adjectives). It is an *intrinsic* property of these terms to be combined in just certain ways to allow a *constructio*; they are all construc*tables* precisely because of this intrinsic property. (b) *Principium efficiens constructionis extrinsecum* is the intellect which "in fact unites the constructibilia prepared and distributed by the modi significandi into constructions and discourse." (FS, 268, n. 37) It is an *extrinsic* property of the intellect that it would ever *effect* signification of an object by a term, but always in accordance with the permissable combinations allowed for intrinsically by the *modi significandi*. (4) The Final cause of *expressing* what is apriorily constructed by the intellect.

13. *Bedeutungslehre* is, accordingly, a "partial region" of logic which treats of meaningful or signifying structures *before* they are used in judgments or asserted in a context and may be assigned a truth-value. The latter would be the domain of logic which in Heidegger's time was called *Urteilslehre* or judgment theory. *Wissenschaftstheorie* was a further domain that distinguished the systematic kinds of assertions or judgments by their subject matters. In any case, signification theory treats grammatically well-formed sentences not only before they are asserted or used in judgments, but before they function in wishes, questions, commands and doubts--i.e., according to Heidegger, before the subject takes a position with one of these complexes (FS, 329).

14. E.g., "The modus activus [significandi] is the act of signification as the achievement [*Leistung*] of consciousness." (FS, 251)

15. E.G., "The act-quality (of the acts of knowing as well as of signification) and the act-matter corresponding to it belong, materially speaking, to different structural regions; in each intentional lived-experience we ought to distinguish the intentional content (Modus passivus) and the [psychologically] real [*reell*] components (Modus activus)." (FS, 262)

16. The extension of the class of objects to cover non-physical objects is found in Heidegger's discussions of the meanings of words like 'blindness' and 'chimera', the problem for "Scotus" of what is signified by *privationes* and *figmenta* (FS, 253-256). Heidegger calls this "Scotus'" "problem of material determinateness". If each word actively signifies, then it must signify some object. All objects that are signified for "Scotus" are ultimately complexes of simply physical objects, as my complex idea of a chimera is ultimately "traceable" (vague) to less complex ideas of a lion, goat and dragon, etc. Since Heidegger, as it

were, relocates the property of passively signifying back several
steps from inherence in the object to inherence in consciousness,
the problem of material-determinateness for him is solved not in
terms of the founding of complex objects on simpler ones, but in
terms of the founding of the abstract matter or noematic-sense of
categorial acts on the more perceptually-oriented matter or sense
of intuitive acts. Though no mention is made here by Heidegger,
in Husserl's theory in LU and *Ideen I*, the ultimate source of all
founding is varyingly in either intuitions of physical particulars
or of sense-data.

   17. Cf.: "It is peculiar that Duns Scotus refers to the
modus essendi which he has just ranked in a line with the *modi
intelligendi* and *significandi passivi*, again in characterizing the
active modi, though without expressly distinguishing a *modus
essendi activus* and *passivus*. We might attribute this to Scotus'
never becoming finally clear about the *modus essendi*, since, on
the one hand, he explains it [the *modus essendi*] as absolute
objective reality, but with this nevertheless does not fail to note
that even it [absolute objective reality] stands under a specific
ratio, namely, that of existence, and so comes close to being a
formal determination, to which an act-character must correspond.
The *modus intelligendi activus* belongs in the region of consciousness, in fact that of the *knowing* consciousness, while the *modus
significandi activus* is to be ordered in the region of "expressions." Now insofar as the acts of knowing, signifying, and that
in which the immediate givenness actually comes into consciousness, are distinguished in each case according to their *ratio*, the
sense of their achievement,...they must also be kept asunder
formally speaking. (FS, 261-262)

   18. This move parallels Husserl's claim in LU VI that all
"categorial" acts (we might say, of reason and understanding) are
based on "intuitive" or "fulfilled" intentional acts. Unfortunately,
the scope of this essay does not permit any discussion of
Husserl's unique unpacking of this traditional Kantian distinction
in terms of "empty" and "fulfilled" acts. Cf. further, R.
Sokolowski, *Husserlian Meditations* (Evanston, Ill.: Northwestern
University Press, 1974).

   19. Cf. FS, 346: "Even as the most general determinations of
objects, which in terms of their content we may view as faded
[*verblasst*], the reflexive categories, are not to be completely
understood without some relation to the *judgment* which constitutes objectivity; this means that a *merely* "objective" universal
theory of objects without regard to the "subjective" side remains

necessarily incomplete. Consequently, every difference is certainly a difference of what is objective, but nevertheless once again *as a cognized, judged* difference. The reason for a manifold of domains of *Geltung* within the categories lies *primarily, though not exclusively*, with the manifold regions of objects, which in each case condition a structured form for forming judgments corresponding to them [the regions of objects], [and] on the basis of which the categories can be 'read off' for the first time in their full content."

20. Cf. Rickert's *Gegenstand der Erkenntnis* (Tübingen: J.C.B. Mohr, 1921), p. ix: "There is still 'another world' than the immanent real one, and in fact it lies in the sphere of value, or it faces us as an *Ought* which may never be reduced to any existent....Hence, we arrive at two worlds, an existing one and one that holds-true." We should note that, after this forward to the third edition, in the forward to the fourth and fifth ones, Rickert no longer finds it necessary to restrict the sense of being to spatio-temporal existence. The third edition (1915) is the one, however, Heidegger was probably working with in 1915-1916 when he prepared the "Scotus"-work.

21. Of note are Lask's two major works, *Die Logik der Philosophie und die Kategorienlehre: Eine Studie über den Herrschaftsbereich der logischen Form* (Tübingen: J.C.B. Mohr, 1911): and in this regard especially his, *Die Lehre vom Urteil*, in *Gesammelte Schriften II*, ed. by Eugen Herrigel (Tübingen: J.C.B. Mohr, 1923). For a full discussion of Lask's judgment-theoretical disputes with Rickert and their influence on Heidegger, cf. Konrad Hobe, "Zwischen Rickert und Heidegger: Versuch eines Perspektiven des Denkens von Emil Lask," *Philosophisches Jahrbuch* 78 (1971), 360-376; also R.M. Stewart, *Psychologism, SINN and URTEIL in the Early Writings of Heidegger*, Unpublished Dissertation, Syracuse University, 1977, Ch. 4.

22. Cf. Caputo's 2nd-mentioned article for an excellent discussion of these themes.

23. For more on Rickert's philosophy of science and history, see his *Grenzen der naturwissenschaftlichen Begriffsbildung* (Tübingen: 1931).

24. I have derived these remarks from several sources: the Preface to FS, XI; Pöggeler's discussion, *op. cit.,* 35-46; Lehmann, *art. cit.,* 341ff.; SZ, 397ff. (449ff.).

25. *Unterwegs zur Sprache* (Pfullingen: Neske Verlag, 1959), 93: "For, as Thou beginnest, so shallt Thou remain."

2. THEODORE J. KISIEL

HEIDEGGER'S EARLY LECTURE COURSES

The latest prospectus on Heidegger's *Gesamtausgabe* (Collected Edition) from the Klostermann publishing house, dated April 1984, brings the welcome news that the majority of his lecture courses during his early Freiburg period (1919–23) as a docent and Husserl's assistant will be published. This decision, which was first posed publicly by the first prospectus of the Collected Edition in the Fall of 1974 and so has been pending for a decade, can only be lauded. The importance of these earliest courses cannot be overestimated for a number of reasons, especially for a philosopher who has repeatedly stressed that his thought is *in via* and that the (sometimes erratic) steps along this way--especially the lecture courses, which accordingly are being published before other still unpublished works by Heidegger--are just as important as the endpoint reached by that thought.

And this particular stretch of Heidegger's *Denkweg* is doubly important inasmuch as these early courses reveal the chronology as well as substance of the very first steps taken by Heidegger toward his magnum opus, *Being and Time* (1927), still regarded by some as the most important single book in 20th century philosophy. Heidegger himself has noted that at least two of the leitmotifs of BT, the environing world and the "hermeneutic of facticity," were the subjects of courses since the winter semester of 1919–1920.[1] It is therefore fortunate that three of the four courses Heidegger held in 1919, including this last course of the year, are among the ones to be published.

For 1919 must be regarded as the pivotal year in Heidegger's development of the ideas which resulted in BT. Simply by consulting Richardson's list of Heidegger's course titles constructed from university catalogues, we note the first appearance of the term 'phenomenology' in that year, and its recurrence in

22

course and seminar titles until 1929.[2] It seems that, after a two-year absence from the university for a stint in the army, Heidegger came back from the front philosophically transformed, an enthusiastic proponent of phenomenology (perhaps naturally, as Husserl's assistant) but, as we shall see, already bent on taking it in his own new direction. The titles of Heidegger's lecture courses before 1919 bear strictly on themes out of the history of philosophy.

But the evidence of university catalogues must be viewed with caution in trying to establish the public record of Heidegger's teaching career. Typically, an instructor must report his course titles early in the previous semester for publication in the university catalogue. He then may later, perhaps as late as the opening day of the following semester, announce a different title to students. This occurred rather frequently with Heidegger. Or in some cases, sometimes for unknown reasons, the announced course was not held at all and no other course took its place. This seems to be the case for the course on "Skepticism in Ancient Philosophy" announced for WS 1922–23. Another such example: In every semester of Heidegger's army service, when he was clearly not at the university, the Freiburg University catalogues announce a course title before his name: SS 1917: "Hegel"; WS 1917–18: "Plato"; SS 1918 *and* WS 1918–19: "Lotze und die Entwicklung der modernen Logik." This was always followed by the "explanation" in parenthesis that he was a "serviceman" (*Kriegsteilnehmer*), so there may have been some bureaucratic or propaganda reason for this listing, which included other instructors as well. It is to be noted once again that all these titles are taken exclusively from the history of philosophy, like the first courses from 1915 to 1917, which presumably were held. But nothing appears to be known about the content of these earliest courses; no original manuscript of them apparently exists: no plans to publish them have been indicated.

This brings us back to 1919. If we supplement the public record of university catalogues with some more substantive internal evidence, namely, extant student notes of Heidegger's lecture courses, we discover that the *very first* lecture course given by Heidegger in that year was in an extraordinary "war emergency semester" (*Kriegsnotsemester* = KNS) from February to April 1919, designed to give war veterans an early start in their interrupted studies. The announced title for Heidegger's course for this special semester was "Kant," but the actual title turned out to be "The Idea of Philosophy and the Problem of World

Views." This course as well as this semester is not reported by
Richardson and, to the best of my knowledge, has never been
reported or discussed in the 'literature.' And the latest prospectus does not announce any plans to publish this course, so presumably there are no extant manuscripts of it written by Heidegger himself. In view of this lack, the importance of this course--
in chronology, it is the very first lecture course of Heidegger's
which is accessible to us; in content, it is Heidegger's very first
phenomenological course (its title alludes to Husserl's *Logos*-
essay) in the 'phenomenological decade' (1919-29) which produced
BT--serves to underscore the value of the student notes (commonly called the *Nachschriften*, 'transcripts') at least in filling
such lacunae in Heidegger's *Denkweg*. From the very beginning
of this post-war period, Heidegger's students made a virtual
'commerce' of comparing, exchanging and passing on their notes,
and so refining them to the point where certain drafts over the
years became prized possessions whose loss, say, in the flight
from East to West Germany in the forties and fifties, was deeply
felt. Some students attempted to take down the words of the
lecture almost verbatim in longhand as it was presented, others
tried to summarize as they listened, and still others formulated
succinct paraphrases (perhaps from several students' notes) after
the hour; some students made an 'official' stenographic record in
shorthand and typed it up afterwards in longhand for Heidegger's
perusal, others passed on their notes in the original shorthand.
Extant *Nachschriften* from the phenomenological decade which is
our topic here include those composed by Oskar Becker (at the
time also Husserl's assistant, on the same level as Heidegger!),
Franz Josef Brecht, Walter Bröcker, Fritz Kaufmann, Hans
Loewald, Karl Löwith, Hermann Mörchen, Simon Moser, Fritz
Neumann, and Helene Weiss.

This 'commerce' in the circulation of transcripts is so well-
known that Heidegger himself alludes to it in his "A Dialogue on
Language between a Japanese and an Inquirer." In this perhaps
quasi-fictional dialogue, the Japanese visitor tentatively recalls a
course given by Heidegger (the 'Inquirer') in 1921 entitled
"Expression and Appearance" attended by his Japanese predeces-
sors, a transcript of which they took to Japan, where it evoked
considerable interest and discussion over the years in the Far
East. Heidegger confirms the title and year, but notes that
"transcripts are really muddy sources"[3] and that the course itself
was most imperfect, since it belongs (he later suggests) to his
juvenilia.[4] In response to the stated reason for interest in the

transcript, namely, the characterization of his way of thought as "hermeneutical," Heidegger recalls tentatively that he first used the words "hermeneutics" and "hermeneutical" in a later course, in the summer of 1923, at the time when he began to jot down the first notes for BT.[5]

The course of SS 1923 indeed bears the title "Ontology: Hermeneutics of Facticity." But the term 'hermeneutics', the subject of considerable discussion in this dialogue, already occurs in the course of KNS 1919 as well as WS 1919-20! And the early lecture course whose title "Expression and Appearance" is meticulously analyzed in the dialogue was in point of fact held in SS 1920 and entitled "Phenomenology of Intuition and Expression." And even though Heidegger devotes several concluding hours in this course of 1920 to his hermeneutic predecessor, Wilhelm Dilthey, the term 'hermeneutics' itself does not seem to be mentioned.

What are the reasons for these departures from the facts in a dialogue where, for example, other dates cited by Heidegger out of his curriculum vitae are accurate? Was it simply the old Heidegger's failing memory over his early years coupled by his failure to cross-check his statements against his old manuscripts? Or was Heidegger exercising a bit of irony over the incessant pursuit of the transcripts of his old lecture courses,[6] extending even to the other side of the globe? Or was it some deeper reason, say, a bit of poetic license to suggest a measure of impatience mixed with the patience with which he tolerated this interest in his juvenilia, whereby one can nevertheless "easily be unfairly judged," just as the old Husserl had "generously tolerated" the young Heidegger's penchant for the *Logical Investigations* twenty years after their first appearance, at a time when Husserl himself no longer held this early work "in very high esteem"?[7] Indeed, this allusion to Husserl's attitude toward his work prefaces the entire discussion of transcripts in the dialogue; and when the conversation later returns to the subject of the transcript of the course on "Expression and Appearance," Heidegger first wonders whether the title after all was not "Expression and Meaning"[8]...the same as Husserl's First Logical Investigation! But this is immediately followed by the in-depth analysis of the 'metaphysical' nature of the terms of the title "Expression and Appearance," and not "Expression and Meaning" or the more historically accurate "Intuition and Expression." One therefore cannot disregard the possibility that Heidegger was deliberately exercising artistic liberty in order to introduce the terms (especially 'hermeneutics') he took to be most appropriate for this Dialogue on Language.

Whatever the reasons for this mixture of fact and fiction (or perhaps error), Heidegger himself was not averse to appealing to "careful philological work"[9] in order to get the record of his own development straight, despite the well-known scorn (often misdirected by 'faithful' Heideggerians) for 'philology' and 'scholarship' which he expressed in other contexts. And the indispensable basis for such work is the factual record of that development. It has already been suggested that the *Nachschriften* can play an invaluable and at times even an indispensable role in setting this factual record straight. Two recent examples of just such a function played by the student notes in the editorial work of the Collected Edition can be cited:

1. In her edition of the course of SS 1925 on "History of the Concept of Time", Petra Jaeger notes that Heidegger's handwritten manuscript ends just short of the Second Division of the Main Part of the course, entitled "The Exposition of Time Itself," whereas the transcripts of both Moser and Weiss indicate that the course really ended with two additional days of lectures (July 30 & 31) on the determination of the wholeness of Dasein through the phenomena of death and conscience. These early reflections on the themes of the Second Division of the First Part of BT are therefore included in the published edition of the course from the text of Moser's typescript.[10]

2. The lecture course of SS 1929, "Introduction to Academic Studies," projected as Volume 29 of the Collected Edition since the very first prospectus, disappears from the latest prospectus with the publication of the course of the next semester as "Volume 29/30" and the announcement by its editor that the course for SS 1929 was not only not held but also not even composed into manuscript form. But Heinrich Petzet, in his book on "encounters with Heidegger," describes his very first "unforgettable" encounter with Heidegger in "WS 1928-29" in a course entitled "Introduction to Academic Studies" in which Heidegger interpreted Plato's Allegory of the Cave. And recently, a *Nachschrift* of the course has surfaced, indicating that it was a one-hour course held, after all, in SS 1929.[11] And even if such a transcript gives us only a rough approximation of the contents of the course, such approximations should not be belittled, particularly for a course located at the end of Heidegger's 'phenomenological decade' and so at the very beginning of his 'turn.' If the transcripts are "muddy sources," they can also be 'reasonable

approximations'; and when nothing else is available, they become indispensable sources. An interesting aspect of this (once again extant) course is that its title harks back to that of SS 1919, "On the Essence of the University and Academic Studies," exactly ten years earlier, just as the course of the previous WS 1928-29, "Introduction to Philosophy," in title and in content recalls the course for KNS 1919. In fact, the issues of all these courses tend to return to the same pivotal question, namely, whether philosophy should be regarded as a science. At the beginning of this phenomological decade, Heidegger sides with Husserl and responds affirmatively to this question. But at the end, philosophy becomes explicit existing out of the ground, the very "happening" of transcendence, and so more primordial than any science and world view.[12]

There are in fact other changes in the factual record which are dictated by the evidence of extant student notes. For example, Heidegger's first course at Marburg in WS 1923-24, generally announced as "The Beginning of Modern Philosophy," (Richardson adds the subtitle "Descartes-Interpretation"), according to student notes actually bears the title "Introduction to Phenomenological Research." Though the course concludes with an interpretation of some of Descrates' texts, it first discusses phenomenology in general and Husserl in particular.

Then there is the case of the Aristotle-Interpretation sent by Heidegger to Natorp in 1922 in support of his application for a professorship at the University of Marburg. Was this transcript, which has been reported as lost, the manuscript of Heidegger's course on Aristotle in SS 1922? If we accept Gadamer's account of the content of the manuscript sent to Natorp[13] and compare it with extant Nachschriften of the course, then it is not: While Paul and Luther are mentioned briefly in conjunction with the classic hermeneutic principle of "determining the dark in terms of the bright," while the Vorgriff of a "hermeneutic situation" is at times invoked in interpreting an Aristotelian text and Augustine is mentioned once fleetingly, August Biel and Peter Lombard are not. So the lecture course does not speak "above all" of the young Luther and Augustine and does not address itself to theological questions "from the very beginning," but is basically a novel translation-paraphrase and exegesis of texts selected primarily from Aristotle's Metaphysics and Physics. What Gadamer describes might well be some sort of hybrid composite of the three courses of 1921-1922, two on Aristotle and one on "Augustine and Neoplatonism."[14] But the evidence for this is inconclusive without Heidegger's original manuscript. And this apparently is not available, since there are no plans to publish this particular course of SS 1922.

In fact, there are two other courses from the early Freiburg period for which no publication plans have been announced, so what we find in the prospectus is not a *complete* list of these courses. And Richardson's list, useful as it was over the years as a starting point and as a "historical document" untouched in its corrections and emendations after it left Heidegger's hand, is ultimately not a *factual* list of the courses actually held at that time. Cross-checking the university catalogues against extant student notes yields the following factual record of titles (here left in the original German) of lecture courses (*Vorlesungen*) given by Heidegger during his early Freiburg period:

KNS 1919 Die Idee der Philosophie und das Weltanschauungs-
problem (2 hours per week).
SS 1919 Phänomenologie und transzendentale Wertphiloso-
phie (1 hour).
SS 1919 Ueber das Wesen der Universität und des akade-
mischen Studiums (1 hour).
WS 1919-20 Grundprobleme der Phänomenologie (2 hours).
SS 1920 Phänomenologie   der   Anschauung   und   des
Ausdrucks: Theorie   der   philosophischen
Begriffsbildung (2 hours).
WS 1920-21 Einleitung in die Phänomenologie der Religion (2
hours).
SS 1921 Augustinus und der Neoplatonismus (announced as
3 hours).
WS 1921-22 Phänomenologische Interpretationen zu Aristoteles:
Einleitung in die phänomenologische Forschung
(2 hours).
SS 1922 Phänomenologische Interpretation ausgewahlter Ab-
handlungen des Aristoteles zur Ontologie und
Logik (4 hours).
SS 1923 Ontologie: Hermeneutik der Faktizität (1 hour).

Some comments on this list: 1. The course of KNS 1919 may have had the simpler title "Idee der Philosophie und Weltanschauung."

2. There is a strong indication that the announced two-hour course for WS 1919-20 entitled "Die philosophischen Grundlagen der mittelalterlichen Mystik" (not listed above) was not held, not

only because no extant transcripts of it have been reported, but also because of the extenuating fact that the course listed above for this semester, originally announced as a one-hour course under the title "Ausgewahlte Probleme der reinen Phänomenologie," was given on the two weekdays (Tuesdays and Fridays) originally specified for the course on "medieval mysticism," instead of on the announced Wednesdays.

    3. For SS 1922, the student notes give an abbreviated title, roughly "Phänomenologische Interpretationen (Aristoteles)," but the above title announced in the catalogue is a fuller and more accurate description of the contents of the course.

    4. No extant transcripts of the two-hour course announced in WS 1922-23 on "Skepticism in Ancient Philosophy" have been reported, so there is some question as to whether it was held. Moreover, no plans to publish it have been announced.

    5. The course of SS 1923, already identified as the first overt step toward the terminology and problems of BT, was only a one-hour course. Dates in student notes indicate that the class was held only on Wednesdays instead of the announced Monday-Thursday schedule.

    6. Finally, though student notes exist for KNS 1919, WS 1920-21 and SS 1922, no plans to publish these courses have been announced, so presumably Heidegger's original handwritten manuscript for them, prerequisite for an "Ausgabe letzter Hand," is not available.

IS THE RETURN TO ORIGINS SCIENTIFIC OR 'HISTORICAL'?

    A glance at the titles of these early Freiburg courses suggests themes so varied that it is at first difficult to discern a guiding thread running through them: philosophy, phenomenology, religion, Aristotle, ontology, hermeneutics. Some scholars who have had occasion to study the contents of some of these courses have tended to underscore the overriding importance of religion or theology (Pöggeler, Gadamer) and Aristotle (Sheehan) in these early years. These judgments might well be premature. There is at least an obvious thematic candidate in the above inventory which must first be thoroughly explored as the overriding concern of these early years, namely, phenomenology, which for Heidegger *is* philosophy, ontology, hermeneutics. This is amply supported by Heidegger's own statements about his early years, his "way to phenomenology" and "through phenomenology to thought." It has already been suggested that Heidegger from

1919 on in his lecture courses overtly identified himself with Husserl's program of phenomenology as the way of establishing philosophy as a strict science. His difference with Husserl—again from the start in this his phenomenological decade (1919-29)—was with the issue of just what the 'matter' of such a philosophy was to be. In what follows, I can make only a few preliminary and fragmentary suggestions about this identity and difference by way of a few selective glimpses into the contents of the courses of this decade in chronological sequence. This will at least serve to underscore the importance of these antecedent Freiburg courses for the true understanding of the Marburg courses which follow.

In KNS 1919, Heidegger confronts the hermeneutic situation of philosophy as he then finds it and locates his approach with respect to philosophy understood as a doctrine of world views (Dilthey) and philosophy regarded as primal science (Aristotle, Husserl). This involved coming to terms then with the Kantian 'transcendental philosophy of values' (Windelband, Rickert) as a kind of middle position: philosophy as a system of values which provides the scientific means for personally developing one's own world view. Heidegger from the start takes a radical approach which breaks with long-standing tradition: philosophy has absolutely nothing to do with world view. It is a primal science which is radically different from all other sciences and is capable of demonstrating itself. With this 'circular' sense of philosophy as his guiding presupposition, Heidegger haltingly works his way through historically available solutions towards trying to define the *matter* of philosophy. Is it really a matter? *How* is it *given*? What does "there is something" (*es gibt etwas*) mean? Thus already in 1919 we find the coining of impersonal sentences in the tradition of Meister Eckhart's *es west* (it essences), like *es wertet* (it values) and *es weltet* (it worlds)! There is even a passing reference to Plato's sense of *aletheia* as the "non-concealed." Experiences from the environing world (*Umwelt*) are already drawn upon to get beyond the quasi-theoretical experience of perception that Husserl made paradigmatic. Lask was the first to see this phenomenological problem of the theoretization of experience without however finding a non-theoretical solution for it. Theoretization de-signifies, de-historicizes, unlives and unworlds our most original experiences. Philosophy's radical quest for a pretheoretical something, not only a worldly but also a preworldly something, makes the primal science at once a supratheoretical science. Philosophy must counter the theoretical tendency of other sciences to unlive the world and replace it with concepts by formulating "recepts" (*Rückgriffe*) which root back in the life-contexts underlying the sciences. The primal sense of this pre-theoretical preworldly something must be seen phenomenally, i.e., purely intuitively. We must learn how to experientially live such lived experiences in their motivations and tendencies. In short, we must come to understand life. For life is not irrational, it is understandable through and through. Phenomenological intuition as the living of primal experiences is *hermeneutic* intuition.

The goal of phenomenology is therefore the investigative re-turn to origins in life itself, not only theoretical but also prac-tical and factical life contexts. It is therefore the science of origins. It is at once the original science, and so not a science at all in the ordinary sense of the word. Thus already in these early years, Heidegger has occasion to assert paradoxically that philosophy is neither a science nor a world view.

But of greater concern in these early years is the deepening definition and shifting formulation of the original domain of phe-nomenology: life-contexts (KNS 1919), situations (SS 1919), factic life-experiences (WS 1919-20)[15]. The basic form of a life-context is a motivation, a situation has the character of tendency, factic life is a vital context of tendencies. These tendencies flow from the self-world where factic life finds its center of gravity, such that the with-world and environing world become functionally dependent upon it. The peaking of factic life in the self-world is already there unaccentuated in the environing world. Everything that happens in the world is defined by the situational character of self-living, its rhythm and style, its habitus in experiencing the world. The fact that factic life can be displaced in its center of gravity into the self-world is not something that happens intentionally, but is rather implicit in the course of life itself. In fact, the very possbility of phenomenology as primal science depends on this possibility. For the many-splendored and some-times chaotic mixture of life-experiences finds its unity in the life of the self. As with any science, preparing the experiential ground for phenomenology and from this shaping the domain of its subject matter and then developing the typical structural con-cepts of its concrete logic are all carried out in and through the factic tendencies of life, and these are always accompanied by a centering in the self-world. On the one hand, we have the idea of phenomenology as an original science, its tendency toward be-ing a science of origins; but on the other hand, this tendency is reciprocally related and motivated by its respective subject

matter. What is that? Not factic life per se but factic life as
originating. Phenomenology wants to find the origin of factic life.
It therefore seeks out the more fundamental situations in which
the totality of life clearly emerges and as it were expresses itself. These are the situations in which I "have myself" explicitly,
not as an object, but in the process of achieving a certain familiarity of life with itself. Phenomenology is nothing other than this
intuitive "moving with" life itself, where it speaks to itself in its
own language through the meaningful contexts that it weaves.
This is the dialectic that philosophy seeks to express, its "dia-
hermeneutics."

And so in SS 1923, the reciprocal relationship between philosophy and its subject matter is entitled "hermeneutics of facticity." It is this formula rather than its equivalent, "ontology of
Dasein," which expresses this central correlation of philosophy as
primal science in the closest possible way. In fact, the "hermeneutics of facticity" is as much a pleonasm as the "botany of
plant." Facticity characterizes the being of "our" "own" Dasein;
it always indicates this Dasein in its instantiation and particu-
larity, insofar as it is 'there' in its being, that is, in the how of
its innermost being and not as an object of an intuition. It is
from and through the 'how' that the character of the being of
Dasein is of itself articulated. Facticity therefore refers to the
self-expression of the being of Dasein.

Hermeneutics is therefore merely the explicit continuation of
the self-explication of facticity. It simply magnifies the possibility
of understanding already inherent in Dasein itself, and so enables
Dasein to be attentive and alert to itself and its ownmost possibilities. It serves to awaken Dasein from its dogmatic slumbers
and so to overcome a certain potential of self-alienation also
inherent in Dasein.

The course of SS 1923 is also noteworthy in Heidegger's
Denkweg for the brief formal introduction of the term *Existenz*.
(It appears in previous courses in its normal non-specific sense
and in the Jaspers review-article of 1921 in conjunction with Jaspers' use of the term.) Facticity is to be interpreted on the basis
of its *Existenz*, that is, on the basis of the authentic possibility
which every Dasein as such is. Such an interpretation of factic
life develops those special concepts called the existentials. But
even though existentials are distinguished from categories and the
source of the term in Kierkegaard is acknowledged, nothing is
really made of this terminology in 1923, and Heidegger continues
to speak of categories for what would later be called existentials.

The same sort of allusion to *Existenz* and the *Existenziale* occurs in the following semester (on the first day in 1924 after the Christmas break), once again only briefly and without follow-through. But this time *Existenz* is located in the third moment of the hermeneutic situation of his interpretation of Dasein (which is the prepossession, *Vorhabe*) in its Being (which is the preview, *Vorsicht*). *Existenz* is now the preconception (*Vorgriff*) according to which what is already had and seen is addressed and considered and so conceptually grasped and articulated.

The same peculiar aversion to the terminology of *Existenz* persists through to the first identifiable draft of BT in the course of SS 1925. (This observation applies also to the much-discussed *Urform* of BT, the lecture on "The Concept of Time" which Heidegger delivered to the Marburg Theological Faculty on Friday, July 25, 1924,[16] where it is once mentioned in passing, without terminological elucidation, that Dasein is "truly existent" in "forerunning" its own death.) As I noted in the Translator's Introduction to this course, *all* the references in the published version of this 1925 course to the formal terminology of *Existenz*, e.g. "existential-ontological" and "ontic-existentiell," are later handwritten changes and marginal comments superimposed upon the stenographic typescript of the lecture course "aus letzter Hand Heideggers," apparently in the process of drafting the final version of BT (probably in early 1926).[17] In view of how completely the terminology of 'existence' dominated BT itself, it is at least odd, if not surprising, that Heidegger hesitated in applying this terminology for several years, right up to the threshold of the final composition of BT.

One problem was, of course, the traditional connotations encumbering the term 'existence,' as Heidegger was soon to experience from his critics, despite the careful distinctions he attempted to draw around the use of his term. Accordingly, for the first time in SS 1925, Heidegger instead introduces the verbal and sometimes clumsy nominative but clearly ontological term "Zusein" (to-be) to characterize the very being of Dasein. This formulation continues to appear in various forms in BT, where e.g. the existentials are still sometimes called *Weisen zu sein* (ways to be). But it becomes as it were a recessive term in Heidegger's *magnum opus*, since it is overpowered there by its much more pervasive replacement term, 'existence.' One can only speculate on the impact Heidegger's most famous book might have had if he had consistently retained this more precisely ontological but cumbersome non-existentialist phrase to express the being of Dasein.

For the term *Zu-sein* has the advantage of being expressly ontological and temporal at the same time and in a particularly telling grammatical way, so that it is much more evident, for example, from the expression 'ways to be' that the so-called 'existentials' are how-categories rather than what-categories. In his very first introduction of this formulation of Dasein's basic character in SS 1925, Heidegger speaks of Dasein as "an entity which *is* in each instance *to be* it [that is, Da-sein] in my own way."[18] The phrase "is to be" is especially telling, since it is a common grammatical way of expressing at once the modals of necessity and possibility; it is thus particularly pregnant in expressing the intricate relationship between the necessity of facticity and the possibility of choice that Heidegger wishes to convey. "That it is and has to be" and that it either can be itself or not underscores the special kind of thatness (*existentia*, factuality) that belongs to Dasein in contradistinction to things.[19]

And the temporal connotations of *Zu-sein* relate back to the discussions in 1919 of the dynamic tendencies of a human situation and how such tendencies are fulfilled (what Heidegger in 1919 called the *Vollzugssinn*, later the *Zeitigungssinn*, of a situation). By 1923, the temporality of Dasein is a central theme for the ontology projected by Heidegger. This culminates in the temporal interpretation of being in SS 1927 in terms of the ecstatic projections of Dasein toward the horizons of time and their schemata. But the course also brings Heidegger's most extreme statements on phenomenology as a scientific philosophy, here more specifically as the science of being.

In what appears to be a deliberate retrieve of the guiding ideas of 1919--the course of 1927 is even titled "The Basic Problems of Phenomenology," like the course of WS 1919-20--Heidegger begins with a discussion of philosophy as science pure and simple versus philosophy as world view. And although Heidegger here continues to insist that the positive sciences and philosophy are absolutely different--the difference in object between a *positum* and a 'nothing'--he now concedes (perhaps under the influence of his intensive reading of Kant during this period) that philosophy must proceed in a way analogous to the positive sciences in order to achieve its scientificity: Just as a positive science objectifies its entity by projecting it upon the horizon of being by which it is understood, so must philosophy objectify being itself by projecting it upon its horizon of understandability, upon something which is beyond being itself, namely time, which allows us to conceptualize being. Thus philosophy (phenomenology, ontology), in contradistinction to the positive sciences, must become a temporal science.[20]

But a year later, philosophy is neither a science nor a world view, but a certain way of existing which provides the basis for both science and world view. The full climax of this upshot of a decade of development is found in the first of the later Freiburg lecture courses in WS 1928-29 entitled "Introduction of Philosophy." Here, Heidegger confronts philosophy in great depth and detail with both science and world view and has them yield to a third relationship: philosophy and history, or more precisely, philosophy and the very happening of Dasein. The talk here is no longer of philosophy and its object, but of philosophizing as explicit transcending, the process of letting transcendence happen from and in its ground, that is, of bringing the very essence of being human into movement. Here, philosophy is not a science not out of lack but out of superabundance, because it is something more original than science. Terms like primal science, original science, strict science become misleading in this context, in part because they make us think of particular sciences like mathematics and physics, which radically contradict the very essence of philosophy.

The seeds for this relationship between philosophy and history were in fact already sown in the courses of 1919. History was then also not understood from the historical sciences, but as an intensification of life, a familiarity with life which comes from living it in its fullness, a co-movement with the very vitality of life which it was incumbent upon phenomenology to establish in order to shape the concepts which would describe that life.

Accordingly, it is far more than the trite title of the first of the later Freiburg lecture courses that prompts us to recall the first of the early post-war Freiburg courses, "The Idea of Philosophy and World View." In this vein, Heidegger was fond of citing a line from Hölderlin's *Der Rhein*: "As you began, so will you remain."

ACKNOWLEDGMENTS

Material for this article was gathered during the tenure of grants from the German Academic Exchange Service (DADD) and the National Endowment for the Humanities. For assistance in realizing this aim, I also wish to express my gratitude to Walter Biemel, Hartmut Buchner, Otto Pöggeler, Andre Schuwer, Joachim W. Storck, and Ernst Tugendhat.

NOTES

1. Martin Heidegger, *Being and Time*, translated by John Macquarrie and Edward Robinson (New York: Harper & Row, 1962), p. 490. Heidegger's note appears on page 72 in the German, or in the common notion, SZ, p. 72. Hereafter I shall use this notation in referring to *Being and Time* (= BT), where the German pagination appears in the margin.

The phrase "hermeneutic of facticity" actually does not appear in Heidegger's lecture courses until 1923, but both "hermeneutic" and "factic life" can be found as early as 1919. Thus Heidegger's memory in 1927 is more accurate than it is three decades later, when he recalls that the word 'hermeneutics' was first used by him in the summer course of 1923 (cf. note 5 below).

2. William J. Richardson, S.J., *Heidegger: Through Phenomenology to Thought* (The Hague: Nijhoff, 1963), pp. 663-66. The courses on Hegel's *Phenomenology of Spirit* are not taken into account here. As in Richardson, hereafter Summer Semester = SS and Winter Semester = WS.

3. Martin Heidegger, *Unterwegs zur Sprache* (Pfullingen: Neske, 1959), p. 91. English translation by Peter Hertz, *On the Way to Language* (New York: Harper & Row, 1971), p. 6.

4. *Ibid.*, p. 128(35).

5. *Ibid.*, p. 95(9). *Niederschriften* is here better translated as "notes" for rather than "drafts" of BT. The first extant draft of BT is probably the course of SS 1925. The course of SS 1923 only sketches a few of the conceptual structures which later provide the framework for BT.

6. Otto Pöggeler, *Der Denkweg Martin Heideggers* (Pfullingen: Neske, 1931), "Nachwort zur zweiten Auflage," p. 351.

7. *Unterwegs zur Sprache*, pp. 128, 90-1(35,5). In fact, in the very same words, Pöggeler (*Ibid.*), relating his own experience of working closely with Heidegger in the late fifties, mentions that the old thinker "generously tolerated" his own interest in the early lecture courses at Freiburg.

8. *Ibid.*, p. 128(34).

9. Pöggeler, *op. cit.*, p. 353.

10. Martin Heidegger, *Prolegomena zur Geschichte des Zeitbegriffs*, Volume 20 of the *Gesamtausgabe*, edited by Petra Jaeger (Frankfurt: Klostermann, 1979), Editor's Epilogue, p. 446. English translation by Theodore Kisiel, *History of the Concept of Time: Prolegomena* (Bloomington: Indiana University Press, 1985),

p. 323.

11. Martin Heidegger, *Die Grundbegriffe der Metaphysik*, Volume 29-30 of the *Gesamtausgabe*, edited by Friedrich-Wilhelm von Hermann (Frankfurt: Klostermann, 1983), p. 537; Heinrich Wiegand Petzet, *Auf einen Stern Zugehen: Begegnungen und Gespräche mit Martin Heidegger 1929-1976* (Frankfurt: Societäts-Verlag, 1983), p. 16; Otto Pöggeler, "Den Führer führen? Heidegger und kein Ende," *Philosophische Rundschau* 32 (1985) 26-27, esp. p. 33.

12. Cf. the course of SS 1928, published as Martin Heidegger, *Metaphysische Anfangsgründe der Logik im Ausgang von Leibniz*, Volume 26 of the *Gesamtausgabe*, edited by Klaus Held (Frankfurt: Klostermann, 1978), pp. 231, 275, 286. English translation by Michael Heim, *The Metaphysical Foundations of Logic* (Bloomington: Indiana University Press, 1984), pp. 180, 212, 221.

13. Hans-Georg Gadamer, *Philosophische Lehrjahre: Eine Rückschau* (Frankfurt: Klostermann, 1977), pp. 212, 24; *Kleine Schriften I: Philosophie, Hermeneutik* (Tübingen: Mohr, 1967), p. 84.

14. According to Sheehan, "sometime between September 22 and October 30, 1922, Heidegger had his brother Fritz type up two copies of a forty-page manuscript [the length reported by Gadamer, *Ibid.*, p. 212] which pulled together the major themes and directions of his teaching over the past three years." Thomas Sheehan, "Heidegger's Early Years: Fragments for a Philosophical Biography," *Heidegger: The Man and the Thinker*, ed. T. Sheehan (Chicago: Precedent, 1981), p. 11. But if Gadamer lost the copy he received from Natorp "in the bombings of Leipzig," (p. 12), where is the copy that Heidegger kept for himself? (According to a recent conversation with Gadamer, the manuscript bore the title "The Hermeneutic Situation.")

15. A brief synopsis of the content of this course is to be found in Pöggeler, *Der Denkweg Martin Heideggers*, pp. 27-28.

16. Gadamer, *Kleine Schriften I*, p. 82. For a summary of this lecture, cf. Oskar Becker, "Mathematische Existenz," *Jahrbuch für Philosophie und phänomenologische Forschung*, 8 (1927), esp. pp. 660-66; Thomas J. Sheehan, "The 'Original Form' of *Sein und Zeit*: Heidegger's *Der Begriff der Zeit* (1924)," *Journal of the British Society for Phenomenology* 10 (1979), pp. 78-83. It is important for our purposes to note that Sheehan here misleadingly translates "Dasein" as "existence" without explicitly informing the reader.

17. Cf. note 10. Unfortunately, the Translator's Introduction originally intended for publication with the translation of this course had to be published elsewhere because of restrictions upon the length and content of prefatory remarks imposed by Heidegger's literary executors. Accordingly, cf. Theodore Kisiel, "On the Way to *Being and Time*: Introduction to the Translation of Heidegger's *Prolegomena zur Geschichte des Zeitbegriffs*," *Research in Phenomenology* 15 (1985), for a discussion of further evidence of drafting in the direction of BT superimposed upon Simon Moser's stenographic typescript, like the initial introduction, later handwritten insertions, and the resulting shift in meaning of the term *Bewandtnis*.

A clear specific example of such drafting is Heidegger's handwritten insertion of a phrase (italicized here) in a sentence of Moser's typescript of the course (German ed., p. 338; Eng. tr., p. 245) which brings the sentence more in line with its parallel in BT (SZ, 127, lines 4 & 5): "This Anyone, who is no one in particular and 'all' are, *although not as a sum*, dictates the mode of being of everyday Dasein." Unfortunately, the German editor, confronted with Heidegger's admittedly difficult handwriting here, did not consult *Sein und Zeit* for assistance and misread *Summe* as *Sein*. A comparison with *Sein und Zeit* here (and in numerous other places) would have also assisted the editor enormously in paragraphing the text of the course (the manuscripts are insufficiently paragraphed) more "according to the wishes of Martin Heidegger" rather than relying solely on her own insights.

Apparently, all six Moser-transcripts in Heidegger's possession during these years pose such problems of editorial exegesis: cf. Friedrich Wilhelm von Herrmann, "Die Edition der Vorlesungen Heideggers in seiner Gesamtausgabe letzter Hand," *Freiburger Universitätsblatter* 78 (December 1982), esp. p. 88. Each editor is therefore called upon to decide which handwritten emendations to include and which not. (One inappropriate inclusion from such handwritten remarks in the German text of SS 1925 is on p. 226: cf. my note to the English translation, p. 167. One serious omission from the summary of a previous lecture hour is the clear distinction between the 'categorial' and the 'categories', which would have appeared in the German text on p. 64: cf. my note to the English translation on p. 48.) Editorial responsibility dictates acknowledgment of the *interpretation* (and the consequent possibility of misinterpretation) that already occurs in the initial editing of these texts, in the numerous minor and sometimes major decisions with regard to, for example, the "expansions," "abridg-ments" and "transpositions" that Heidegger himself foresaw as necessary in the editing of his lecture courses. It is therefore incredible to me that a hermeneutically astute philosopher like Heidegger could have even coined the phrase "edition without interpretation" to characterize his *Gesamtaus-gabe*. Much more discerning in this context are the prefatory remarks to this issue of the *Freiburger Universitätsblatter* on the overall topic of "Edition und Interpretation" made by its general editor, Gerhard Neumann, who refers to the recent tradition in German philology which maintains that "edition *is* interpretation" (p. 11, note 6).

18. *Prolegomena*, p. 205(135). The comma after "das ist" is questionable.

19. SZ, 135, 12. For a linguistic analysis of the 'telling' complexity of the Heideggerian term *Zu-sein*, cf. Ernst Tungend-hat, *Selbstbewusstsein und Selbstbestimmung* (Frankfurt: Suhr-kamp, 1979), pp. 36-37, 176-91. Cf. also my *Translator's Intro-duction* for a fuller discussion of the role and development of *Zu-sein* in this course.

20. Martin Heidegger, *Die Grundprobleme der Phänomenolo-gie*, Volume 24 of the *Gesamtausgabe*, edited by Friedrich-Wilhelm von Hermann (Frankfurt: Klostermann, 1975), §§ 1-3, 20b, 22b. English translation by Albert Hofstadter, *The Basic Problems of Phenomenology* (Bloomington: Indiana University Press, 1982).

3. THOMAS SHEEHAN

*HEIDEGGER'S "INTRODUCTION TO THE PHENOMENOLOGY
OF RELIGION," 1920-21**

When Martin Heidegger's *Being and Time* was born in 1927,
its philosophical genealogy was quite obscure. It was dedicated to
Edmund Husserl, but criticized the master throughout. It paraded
its debts to Kierkegaard and Jaspers, but adjudged both thinkers
to be inadequate. It drew upon Aristotle at virtually every turn,
but roundly repudiated his metaphysics of substance. Clearly the
work was a creative mix which at one and the same time appropriated and expropriated the whole philosophical tradition.

But another major influence on the uniqueness of *Being and
Time* lay concealed in the background: Heidegger's reading of St.
Paul and early Christianity. We now know the great extent to
which Aristotle contributed to the themes and vectors of Heidegger's project, and the underlying Kiergaardian themes have
long been discussed. But we are only beginning to gather the
evidence which will allow us to document, rather than just to
surmise, the strong influence which early Christianity exerted on
Being and Time.

The topic is broad and complex, and in this essay I focus
on only one element: the lecture course which the young Privatdozent delivered at Freiburg University in the winter semester of
1920-21. *Einleitung in die Phänomenologie der Religion* ("Introduction to the Phenomenology of Religion"). The monumental *Gesamtausgabe* or Collected Edition of Heidegger's books and lecture
courses, which began to appear in 1975, will probably not include
Heidegger's early lectures at Freiburg between 1919 and 1923,
since Heidegger's own notes for these courses are sketchy and

———————
*This article first appeared in *The Personalist*, 55(1979-
1980), pp. 312-324. Reprinted by permission from the author and
the editors.

often preliminary. But the pity is that we will therefore be denied an insight into this creative period of maturation. However, I have had the opportunity to study some of the transcripts (*Nachschriften*) which Heidegger's students made during the lecture course on phenomenology of religion, and I shall draw upon these documents for the report which follows.

Heidegger's interest in religion was already deep and of long standing before he began this course. We know of the religious convictions of his family, of his own studies towards the Jesuit priesthood as a young man, and of his abiding interest in theology and religion right up until his death and Catholic burial in 1976. To set the stage for his 1920-21 course, I shall begin with some recently gathered evidence of a biographical nature which sheds light on Heidegger's personal religious orientation in the years just prior to his lecture. I will discuss three letters that Edmund Husserl wrote about the young Heidegger from 1917 through 1920.

I

On Monday, October 8, 1917, Edmund Husserl received a letter in Freiburg from his colleague, Paul Natorp of the Philipps University at Marburg. This was the first in a long exchange of letters they would have, until Natorp's death, about the man whom Husserl would call his *Lieblingsschüler* ("favorite student"), Martin Heidegger. But in the fall of 1917 Heidegger was a person whom Husserl did not know well. When Husserl arrived in Freiburg in May of 1916 and met the young scholar, Heidegger had already presented his *Habilitationsschrift* to the philosophy faculty. In September of 1916 Husserl helped Heidegger get the work published, and when the book--on Duns Scotus' doctrine of categories and meaning--came out, Heidegger sent Husserl a copy inscribed, "To Prof. Husserl with most grateful respect." But some months later Heidegger was drafted into the army, and occasions for the two to get closely acquainted were curtailed.[1]

Meanwhile, Natorp was interested in filling a teaching position at Marburg in the history of medieval philosophy. He had read Heidegger's book on Duns Scotus and was writing to ask Husserl if the twenty-eight year old Heidegger would be a suitable candidate. In particular, Natorp wondered "whether one can be really quite sure about any religious [*konfessioneller*] narrowness in him."[2] Husserl drafted his response on the same day he received the letter.

I hasten to respond to your inquiry. Because Dr. Heidegger
is very busy in the war service, I have not had sufficient
opportunity up to this point to become closely acquainted with
him and to form for myself a reliable judgment about his personality and character. In any case I have nothing bad to say
of him.

It is certain that he has confessional ties, because he
stands, so to speak, under the protection of our "Catholic
historian," my colleague Finke. Accordingly last year he was
proposed in committee meetings as a nominee for the chair of
Catholic philosophy in our Philosophy Department—a chair
which we too would have liked to form into a scientific teaching position for the history of medieval philosophy. He was
taken into consideration along with others, at which point
Finke suggested him as an appropriate candidate in terms of
his religious affiliation [*in konfessioneller Hinsicht*].[3]

Husserl goes on to say that Heidegger is too young and not ripe
enough for the professorship at Marburg. His Duns Scotus book
is only a beginner's work (*Erstlingsbuch*), and although Heidegger as a teacher maintains "a secure position on fundamental
questions and methods," Husserl has heard "some very favorable
judgments and also some critical ones—which in any case is connected with the fact that, in order to make headway in the systematic area, he does not give historical lecture courses but systematic ones." Moreover, Heidegger is no longer satisfied with
Rickert, and "he now seeks to come to grips with phenomenological philosophy from within. It seems that he is doing this seriously and with thoroughness. This is all that I can say at this
time." Needless to say, with faint praise like that, Heidegger did
not get the position.

A year and a half later, a second inquiry came to Husserl
from Marburg, this time from the Protestant theologian, Rudolf
Otto. Interestingly, Otto asks not about Heidegger but about a
certain Mr. Oxner, but in his response of March 5, 1919, Husserl
loses no time in getting around to mentioning Heidegger. (This is
not the last of the Otto-Heidegger relationship. Seven years
later, Otto would strongly but unsuccessfully oppose Heidegger's
nomination to succeed Hartmann at Marburg.[4]) Husserl writes
about Heidegger, religion, and Otto's book *The Holy*, as it were,
in one breath.

Mr. Oxner, like his older friend, Dr. Heidegger, was originally a philosophy student of Rickert. Not without difficult inner struggles did both of them gradually open themselves to my suggestions and come closer to me personally. In that same period, both of them underwent radical changes in their basic religious convictions. Truly, both of them are religiously directed personalities. In Heidegger it is the theoretical-philosophical interest which predominates, whereas in Oxner it is the religious....

My philosophical effect has something remarkably revolutionary about it: Protestants become Catholic. Catholics become Protestant....In arch-Catholic Freiburg I do not want to stand out as a corrupter of the youth, as a proselytizer, as an enemy of the Catholic Church. That I am not, I have not exercised the least influence on Heidegger's and Oxner's migration over to the ground of Protestantism, although it can only be very pleasing to me as a "non-dogmatic Protestant" and a free Christian (if one may call himself such when, by "free Christian," he envisages an ideal goal of religious longing and understands it, for his part, as an infinite task). For the rest, I like to have an effect on all sincere men, whether Catholic, Protestant or Jewish.[5]

Husserl goes on to say that Heidegger and Oxner made him aware, in the summer of 1918, of Otto's book, *Das Heilige,* which had a strong impact on him. I will return to this matter at the end of this essay. The point now is to underscore three things: that Heidegger undergoes a radical shift in his religious convictions between 1916 and 1919; that the shift consists at least in a move from Catholicism to Protestantism; and that even in this shift Heidegger's interests remain theoretical-philosophical and not just religious.

These matters re-emerge a year later when we find Husserl again writing to Paul Natorp (February 11, 1920) and referring back to their first letter on Heidegger in 1917.

A thought comes to me very late and disturbs me within. The *Extraordinarius* position in philosophy at Marburg was recently filled and when it was last vacant, Dr. Heidegger was also considered for it, as you graciously informed me in due course. Now it disturbs me that I wrote at that time that Heidegger had entered the faculty here as a "Catholic philosopher," and [I fear] that this could now again cut off consideration of putting him on the list. Allow me to inform you that, although I didn't know it at the time, Heidegger had already freed himself then from dogmatic Catholicism. Soon afterwards he drew all the conclusions and has cut himself off--clearly, energetically and yet tactfully--from the sure and easy career of a "philosopher of the Catholic worldview." In the last two years he has been my most valuable philosophical co-worker. I have the very best impressions of him as an academic teacher and a philosophical thinker, and I place great hopes in him. His seminar meetings are as well attended as my own, and he knows how to fascinate beginners and advanced students at the same time. Moreover, his very famous lecture courses--polished in form and yet quite profound--are very strongly attended (about 100 students). He has worked his way into phenomenology with the greatest energy, and he generally strives to lay the surest foundations for his philosophical thinking. His scholarship is wideranging. A sterling person.

Yet at that time I got stuck in a questionable recommendation which I had no right to make and which I beg will not be taken amiss. But because you yourself, dear colleague, once asked me about him and because at that time I could not give an entirely unambiguous and securely founded judgment as I can today, it was a matter of conscience to make up for it now, especially since I can say at this time that he is one of the most promising young men whom we have to look out for (in his material plight which, however, he bears lightheartedly. Heidegger--to use the Viennese expression--is not a *Raunzer* [whimperer]).[6]

This third letter comes at a turning point in Heidegger's career. It was written precisely at the time when Heidegger was beginning the development that would culminate in *Being and Time*. For in February of 1920, the date of that third letter, Heidegger was engaged in teaching the famous lecture course entitled "Basic Problems of Phenomenology," (*Grundprobleme der Phänomenologie*[7]), in which for the first time he presented his analysis of environment and, in general, his hermeneutics of the facticity of existence. This was also the topic which structured his lecture course in the fall and winter of that year, "Introduction to the Phenomenology of Religion."

II

Heidegger's course, *Einleitung in die Phänomenologie der Religion*, met twice a week, on Tuesdays and Fridays from noon to one o'clock, from the beginning of November, 1920, through February 25, 1921, with a month's recess for Christmas. Here, in schematic fashion, is an outline of the course. (The articulations are my own.)

Part One
(8 lectures)
Introduction to the Phenomenon of Factical Life-Experience

Part Two
(16 lectures)
A Phenomenological Interpretation of Original Christianity in St. Paul's Epistles to the Galatians and Thessalonians
Section One: How Original Christianity is a
Factical Life-Experience
Section Two: How Original Christianity, *as*
Factical Life-Experience, is
Primordial Temporality.

To give an idea of the whole of the lecture in this brief essay, I shall first run through Part One of the course, quickly and somewhat superficially, in order to point out the major divisions of thematic treatment. Then I shall focus on Part Two, specifically on the interpretation of the First and Second Epistles to the Thessalonians. But first, a word on the viewpoint that I bring to bear on this material.

In reading this course some sixty years after it was delivered, one finds a curious phenomenon coming into play, viz., that today's reader has a fuller perspective on it than even Heidegger himself did in 1920. Today we can see it as one step in a series of moves that led to the writing of *Being and Time* in 1923-26, whereas it is difficult to say to what degree, if at all, Heidegger's sights were already set on that latter work when he was teaching this course. Was he interested in St. Paul and the phenomenology of religion for their own sake, or only as a means to working out the fundamental structures of his own project of elaborating the analogically unified meaning of being on the basis of a new understanding of temporality? Although this question cannot be answered here (or perhaps ever), I mention this

curious phenomenon of "foreshortening"--that is, of seeing this
course in the perspective of *Being and Time*, or *Being and Time*
against the background of this course--because for this reader at
least, such foreshortening seems ineluctable. In what follows I
will be discussing this course not only for what it says about the
method and theme of phenomenology of religion, but also for what
it lets us see of the development of *Being and Time*.

The Phenomenon of Factical Life-Experience

Part One of the course introduces a phenomenology dedicated
to recovering what was forgotten by the entire Western tradition
(including Husserl), but which, even if unthematically, was
understood by early Christianity: life in its here-and-now facticity, factical life-experience (*das faktische Leben, die faktische
Lebenserfahrung*). This is a technical term and an awkward one,
which *Being and Time* will replace with the more manageable term
Dasein, "existence." Notice the parallels between Heidegger's
approach to early Christianity and his approach to the Greeks. In
both cases he sees a lived but unthematized level of experience
which was covered over by subsequent ages and which must be
recovered by a "violent" de-construction of the tradition. Note as
well the parallels in the content of these two experiences. In both
cases the experience is a pre-theoretical, pre-rational lived
experience of "self-exceeding," of being drawn out beyond one's
ordinary self-understanding. In considering the pre-Socratics,
Heidegger will spell out this experience in terms of *phusis*, *logos*,
and *aletheia*. In dealing with early Christianity, he spells out the
pre-theoretical element in terms of life-experience, and the self-
exceeding "drawn-out-ness" in terms of facticity. What both cases
share in common is *movement* as a dynamic interplay of presence
and privative absence (pres-ab-sence). The difference between
the two experiences is in this nuance: In early Christianity this
primordial pres-ab-sential movement is understood in terms of
temporality, whereas in early Greek experience, the pres-ab-
sential movement is thematized in terms of *disclosure* or "truth."
Temporality and truth--the two ways of looking at Heidegger's
one and only topic--are rooted respectively in his readings of
early Christianity and of archaic Greece.
    I shall cut through a great deal of material here in order to
engage only two questions which emerge in Part One of the
course: (1) What is factical life-experience? and (2) How does
this phenomenon prescribe a new phenomenological method?

The course opens by asking how a philosopher should carry out an "introduction to philosophy." Such an introduction would seem obliged to treat of the "pre-questions" of philosophy (for example, how philosophy is different from science and religion), after which one supposedly could get on with philosophy itself. But, says Heidegger, the real meaning of philosophizing is *always to stay* with these questions. Indeed, Heidegger would make a virtue of staying with the *Vorfragen*, if only because these prior questions about the meaning of philosophy's conceptuality have not really been raised since Socrates (lecture of January 21, 1921). But the real pre-question of philosophy is the very questionability of life itself, and this he calls "the factical experience of life." To burrow down that far is not only to cut below the distinction between philosophy and science, but also to find the elusive basis for philosophy itself, the ground out of which philosophy grows and to which it must finally return. We recognize here an early statement of what is said in *Being and Time* (German, p. 38): Philosophy as universal phenomenological ontology takes its departure from the hermeneutic of Dasein which ties all philosophical inquiry down to the point whence it arises and to which it returns: existence.

But to step back into factical life-experience entails a complete transformation of philosophy (*eine völlige Umwandlung der Philosophie*). Heidegger distinguishes this transformation from the mere change of viewpoint (*Blickwendung*) required by Husserl's phenomenology. What Heidegger is referring to here—as early as 1920—is what later he calls the "turn" or *Kehre*. Clearly, the "turn" is not the *de facto* shift in language and approach that happened in Heidegger's thought in the thirties. Rather, it was envisioned and built into his project from the beginning, and it meant a radical turn *away from* all philosophizing built on the correlation between man as a stable subject and beingness as the stable presentness of things *and into* the primordial experience of being thrown into nothingness (privative absence) within which things become meaningfully present: the event of pres-ab-sence. The first halting name for this event of (*Ereignis*) is "factical life-experience."

But what is factical life-experience (Dasein)? It is not mere cognitive experience but rather man's overall coming-to-grips with the world (*das Sich-auseinandersetzen des Menschen mit der Welt*). We notice the "phenomenological correlation" operative here and its radical difference from such correlation in Husserl. Thus, if we speak of the "object" of this experience, or better, that

which is had in it (*das Gehabte*) or that which is lived out in
this lived experience (*das Erlebte*), we are not talking about mere
objects for a knowing subject. We are referring, rather, to a
world of meaning in which one lives, the environment in which
one acts, the shared world to which other men individually and
socially belong, and the self-world in which one is concerned
about oneself (*Umwelt, Mitwelt, Selbstwelt*). Here is Heidegger's
transformation of the transcendental correlation of Husserl's
phenomenology: no longer a correlation of cognitive *noesis* and
noema, but rather one of lived experience on the one hand and
modalities of world (i.e., lived meaning) on the other. These
three levels or modalities of world are, Heidegger says, accessible
only in and through factical life-experience. Moreover, factical
experience exhausts the whole range of living. Indifferent and
self-sufficient in itself, it stretches even to the highest things of
experience and deals with them as the factically experienced.

Heidegger's opposition to Husserl's pure ego was not, as is
often claimed, something which he hid from his master and only
shared with students in the classroom. In that regard, a letter
from one Frau Walter of Freiburg to Prof. Pfänder on June 20,
1919--a full year and six months before this lecture course--is
revelatory. At one of the Saturday morning discussions which
Husserl was accustomed to have at his house on Lorettostrasse,
Walter reports, a "campaign...against the pure ego" was recently
launched by the young Dr. Ebbinghaus and followed up by Heidegger. Whereas Ebbinghaus, from his Fichtean and Hegelian
standpoint, entirely dismissed the possibility of an empty,
essence-less pure ego, Heidegger--who also opposed such an
ego--took a mediating position between Ebbinghaus and Husserl.
Walter writes that "he says the primordial Ur-ego would be the
'historical ego' qualified; the pure ego would arise from the
'historical ego' by the repression of all historicity and quality,
but it could only be the subject of material-theoretical acts
[*sachlich-theoretischer Akte*]."[8] From this report it would seem
that Heidegger's attack on the pure ego of Husserl, which he
published in *Being and Time* in 1927 and spoke forth in his
course in 1920, was one which he made known to Husserl from the
beginning. And the crux of the matter concerns the "correlation"
which Heidegger sees between factical existence and its world. He
expressed it clearly and concisely in his 1925-26 course on *Logic:
The Question of Truth*. The correlation is deeper than any intentional relation between consciousness and its objects, it is "truth"
and the primordial opening up of the realm of experience. "If we

can speak this way at all, it [truth] is the relation of existence as existence to its very world, the world-openness of existence, its Being towards the world itself--which world gets opened up, revealed, in and with this Being towards it."[9]

The continuity of the analysis of factical life-experience in 1920 and the later *Being and Time* is further shown when Heidegger points to the unifying characteristic of everything experienced in factical life-experience: *Bedeutsamkeit*, meaningfulness or significance (cf. *Being and Time* § 18). This, of course, does not have any connotations of epistemology or of Rickert's value philosophy. Rather, meaningfulness is *lived through* experientially in what Heidegger calls *Bekümmerung* and *Sich-Bekümmerung*, the earliest names for what *Being and Time* will later call, respectively, *Besorgen* and *Sorge*, concern and care. He goes on: Man in his concern and care is captured by his world, so much so that we must speak of an intrinsic dimension of experience that is a "falling" (*Abfall*) into meaningfulness. In fact, we never experience ourselves in separation from a lived matrix of meaning, but only in such a way that we are caught up in the *Tun und Leiden*, the acting and suffering, of our world. Yet at the same time, if everything we experience has such a world-character, experience is likewise always experience of the self-world (*Selbstwelterfahrung*).

Heidegger summarizes his treatment of factical life-experience in a lengthy Teutonic phrase that in some ways previews the definition of care in *Being and Time*. Factical life-experience is "*Bedeutsamkeitsbekümmerung*" (concern for meaningfulness), a concern that is qualified by four adjectival phrases: (1) *einstellungsmässig*: perhaps, "involved" or "committed"; (2) *abfallend*: "falling"; (3) *bezugsmassig-indifferent*: it cuts through and characterizes all modes of relationality; and (4) *selbstgenügsam*: self-sufficient and underived.

Thus far, a quick sketch of what factical life-experience is. The second question of Part One of the course is now: How does this experience already prescribe its own new method of investigation? That is, how is phenomenology transformed from its Husserlian shape when the basic phenomenon of phenomenology is taken to be factical life-experience?

Stated generally, the major transformation consists in the turn to the "historical." (At this early phase Heidegger uses the word *"historisch,"* but it is clear that he means not the ordinary "historical" but rather what he later would call the *geschichtlich.*") Factical life-experience is intrinsically historical and hence, methodologically, can only be understood from out of an understanding of the historical. He says: *Philosophie ist der Rückgang ins Ursprünglich-historische*. Philosophy is the return to the primordial historical. Here again an early statement of the "turn."

More specifically, Heidegger works out a new direction for the phenomenological method by analyzing the meaning of "experience" (*Erfahrung*) in the phrase "lived experience." Experience can be taken to mean two things: first, that-which-is-experienced *(das Erfahrene)* and, secondly, the experiencing of that-which-is-experienced *(das Erfahren, die Erfahrung)*. (We recognize here again a re-reading of the noetic-noematic correlation.) Indeed, these two are not to be separated as if they were two things, but rather are bound together in the complex structure of what Heidegger calls, in a very broad sense, "intentionality." In their unity they are "the thing itself," the phenomenon. "Phenomenon," therefore, is not just *what* is experienced but as well the mode of experiencing of what is experienced, i.e., the way in which the "what" shows itself, its "how" of appearing. We can already see here that even when Heidegger begins with ordinary everyday experience *(Alltäglichkeit)*, he does not do so within some naive realist perspective that would be in constant danger of falling into a positivism of the obvious, but rather he is always operating within what, in the most general sense, we must call a phenomenological reduction. Husserl once remarked that Heidegger's treatment of the environment was rooted in Husserl's own *Ideas*, paragraph 27, "The World of the Natural Standpoint: Ego and Umwelt." That requires some nuancing. Even though embedded ineluctably within the nitty-gritty of dealing with the world around, factical life-experience as investigated by Heidegger is always already "reduced" (i.e., led back to) its "how" of appearing. Whereas Heidegger's "everydayness" certainly corresponds to Husserl's "natural standpoint," Heidegger's phenomenological investigation, far from forgetting the phenomenological reduction, always moves back from things as immediately given "out there" to a consideration of the mode of the experience of them.

The general correlation of experienced and experiencing is further analyzed into three distinct but inseparable moments of meaning: *Gehaltssinn, Bezugssinn,* and most important of all, *Vollzugssinn*. *Gehaltssinn:* This means the primordial sense had in the content of what is experienced. *Bezugssinn*: This is the relational meaning or sense contained in the primordial "how" or way-in-which of the act of experience *(das Wie des Erfahrenwerdens)*.

Lastly, *Vollzugssinn:* This sense or meaning of the very enactment of experience also deals with the "how," but with this nuance: the "how" or way in which the *Bezugssinn* itself is carried out. For Heidegger this is the core of the phenomenon, and it is to be determined in its time-character. The task of phenomenology is to thematize the very temporal enactment of the event of meaning, and not primarily that-which-it-intends nor the relations established in intentionality. This enactment is what Heidegger means by the "how." Later in the course he says: "All questions in philosophy are basically questions about the 'how' (*Wie*) or, strictly understood, questions about method." And "method" here is to be taken in the Aristotelian sense of *methodos*, "pursuit" of a subject matter (cf. *Physics*, G, 1, 200 b 12f.) and not in the sense of a "technique" It appears that for Heidegger this *methodos* is ultimately nothing other than temporality. Therefore, if one takes factical life-experience (as Heidegger does in this period) as the central phenomenon of all phenomenology, then phenomenology itself is transformed. The generation of temporality (or equally, of "the historical") is both the theme and the method of the new phenomenology.

Before we move into the material of Heidegger's analysis of St. Paul, it is worth pausing over the threefold distinction which he makes within the phenomenon. It is, of course, premature at this stage of Heidegger's development to speak of the full-blown Being-problematic, but in the distinction of content-sense, relation-sense and enactment-sense we may perhaps begin to see a primitive articulation of the later distinction between *das Seiende* (the thematic entity), *Seiendheit* (the beingness of that entity = the Greek *ousia*), and *das Sein selbst* (the event of being itself). For Heidegger, the tradition's "beingness," in its various historical transformations, bespeaks the *mode of presentness* of whatever is meaningfully present; hence, it is the "how" of rationality, although mostly this factor of relation is forgotten or covered over. To retrieve beingness, in all of its forms, as a phenomenon of relationality is to have uncovered the implicit but mostly forgotten phenomenological basis of traditional ontology. But that is only the first step. The further move is the uncovering of the very enactment of such relationality in its time-determined character. And since time always comports a negativity (e.g., the "not yet"), this move is a retrieval of the problem of the nothing in traditional ontology. To move into this area is to turn definitively from all philosophy heretofore (including even the phenomenological reading of beingness as stable presentness to a stable

subject) and to enter the area of man's temporal projectedness
into the "nothing" whence things become meaningfully present. To
the eye which sees the whole of Heidegger's project, the
Vollzugssinn or enactment-sense looks like a first name for
Ereignis.

St. Paul and Original Christianity

Part Two of the course was devoted to a phenomenological
interpretation first and briefly of St. Paul's Epistle to the
Galatians and then of his two Epistles to the Thessalonians.
Heidegger chose the latter because they are the oldest Christian
documents, predating even the Gospels. Their very antiquity, he
hoped, would reveal the original features of Christian religious
lived experience (= Pauline *zoe*, life).

The first problem in all phenomenology is always the question of access to the phenomenon--not that phenomena are not
available, but rather that they are too available and therefore
available in deceptive modes of appearance. How do we get the
phenomenon properly in view? This is the question of the "pre-
grasp" or *Vorgriff*, which Heidegger here defines generally as the
tendencies which already priorly motivate one's stating of the
problem. If, in broad terms, the goal of a phenomenology of
religion is the phenomenological understanding of original
Christian religious lived experience from out of itself, then what
is the proper *Vorgriff* of this phenomenon?

In contemporary philosophy of religion, Heidegger says (and
here he has Otto in mind), a major comprehending pre-grasp
(*Vorstehensvorgriff*) of the phenomenon of religion is had in the
contrast of rational and irrational. But this is quite inadequate.
To place religious phenomena in the category of the irrational, no
matter how broadly one characterizes this, is to explain nothing,
precisely because the contrast lives off the meaning of the
"rational," which itself goes unquestioned or unclarified. Heidegger proposes an alternative pre-grasp, which he admits is only an
hypothesis at this stage, namely, that religious phenomena are to
be approached in terms of factical life-experience and specifically
in terms of the temporality and historicity of that experience. In
a statement written contemporaneously with this course, he delin-
eates what he means by the phenomenon of "the historical": an
authentic "stretch" of existence into its past--not a past which is
like some baggage which existence drags along behind itself, but
a past which is experienced historically, i.e., in such a way that

existence possesses it and itself within the horizon of expectations which it has already projected ahead of itself.[10] "The historical," therefore, is the self, that is, the having-of-oneself by the enacting of one's own existence in historical contexts. This concernful, historically enacted self-having is the "how" of man's being, and: "The phenomenological explication of the 'how' of this enactment of experience according to its fundamental *historical* sense is the decisive task in this whole complex of problems involving the phenomenon of existence."[11] We see how the material of Part One of the course flows into and structures the approach in Part Two. Heidegger's pre-grasp focuses on the enactment (*Vollzug*) of religious experience and specifically on the how (*Wie*) of that enactment. Concretely that means that he will work out what he calls the "*vollzugsgeschichtliche Situation*": the religious situation insofar as this is enacted historically. The pre-grasp of the factually historically enacted situation thus becomes the hermeneutical guiding principle of Heidegger's interpretation of St. Paul.

"Situation," he says, is entirely a phenomenological term and is not to be understood as some kind of natural-spatial thing, nor as an objectifiable historical matrix, nor in terms of some kind of stream or flow of experience. To be sure, it is a kind of movement, but not an *ordnungsmässige* movement—we may interpret: not a movement that can be numbered according to before and after. As a phenomenological phenomenon, "situation" is only accessible through, and can only be elaborated from out of, performed or enacted understanding (*vollzugsmässiges Verstehen*). Here we see already the first attempt to delineate the phenomenon of situation as this will emerge in *Being and Time*. There, in the context of the discussion of conscience and resolve, situation gets defined as the openness that existence is, but specifically this openness insofar as it is opened up in the decision to accept oneself as finite.[12] Here, in the context of early Christian experience, Heidegger's goal is first to work out the religious situation of St. Paul's letters to the Thessalonians—i.e., the factical life-experience that structures that situation—and secondly, to specify that situational experience in terms of its temporality.

The first task is carried out mostly by a reading of the First Epistle to the Thessalonians, whereas the second task is in good measure reserved to Heidegger's considerations of eschatology within both epistles to the Thessalonians.

Heidegger's interpretation of I Thessalonians is one of the earliest and most intriguing examples of his hermeneutic of a text. The orientation is, in the broadest sense, linguistic, but neither philological nor exegetical. The concrete situation of the epistle is designated as "preaching," within which particular words, especially when repeated time and again, take on a special interpretative significance. Heidegger points out that within the first twelve verses of the epistle there is a striking repetition of two sets of terms: on the one hand, various forms of the verb *genesthai* (to have become), and on the other, forms of the verbs *mnaomai* (I remember) and *eido* (I know; infinitive *eidenai*: perfect tense used as present: *oida*). Without detailing here his verse-by-verse reading of these words, I shall simply summarize the import of his interpretation.

Genesthai in all its forms points to the basic state of being of St. Paul and of the Thessalonians, namely, their "already having become" or "already having been" (*Gewordensein*). This is not to be taken in the sense of some past and by-gone event (*das Vergangene*), but rather as the whole of what is already operative and determinative of the Thessalonians' present "now." Their *Gewordensein* is their *jetziges Sein*. Here we recognize one of the earliest statements about what *Being and Time* will call *Faktizität* and ultimately *Gewesenheit*. The first is translated as "facticity," that is, the state of being already in the world of sense and involvement. The second term is more difficult to bring into English. The usual translations emphasize the sense of "having been" ("is-as-having-been"), but Heidegger means us to hear no "past" sense in the word *Gewesenheit* but rather, resonances of "that which is already operative." Perhaps we may translate *Gewesenheit* as "the already-dimension of existence," or more simply as "alreadiness." This dimension, we recall from *Being and Time*, does not lie behind existence but rather structures the current "is" of existence, so much so that Heidegger can say that existence is what it already is.[13] Indeed, "alreadiness" co-structures existence's future insofar as it "does not follow behind existence, but rather always goes before it" (*"geht ihm je schon vorweg"*). This reversal of the place of the so-called "past" in Heidegger's work is one of its most significant--but least remarked--achievements.

But equiprimordial with this alreadiness expressed in forms of *genesthai* there is an *eidenai*-dimension--a kind of "experiential knowledge." St. Paul constantly repeats such phrases as: "I have no need to write to you about such and so, for you *already know*..." (cf. 4:9;5:1). This kind of "knowledge" differs from any ordinary cognition or memory, for it is "comprehension of the situation," and hence is had only from out of the situation, that is, from out of factically lived experience. We recognize here what *Being and Time* will call *"Verstehen"* as *"Seinkönnen"*--which is not to be translated as "understanding" *qua* "potentiality for Being" (a momentously misleading rendering), but rather as "savvy" or "know-how" (*Sich-auskennen*; cf. the Greek *empeiria*) as "knowing one's way around in one's own being." One has this kind of "understanding" by taking risks, trying things out, "projecting" them.

These two moments of existence are always found together, the one as an "adjective" to the other. The *genesthai*-dimension gets specified by St. Paul in terms of affectivity (the text mentions *thlipsis* and *chara*, "tribulation" and "joy")--what *Being and Time* will discuss in terms of "affective disposition" or, less felicitously, "state-of-mind" (*Befindlichkeit*). All knowing-one's-way-around is rooted in the affective disposition in which one already finds oneself. *Genesthai* and *eidenai* (these two terms pointing to the phenomenological roots respectively of "body" and "mind") are early signposts leading towards the major existential structures of *Being and Time*. viz., *Befindlichkeit* and *Verstehen*, which in turn point to the temporal moments of *Gewesenheit* and *Zukünftigkeit*. So we turn to Heidegger's second thesis, namely, that original Christian experience generates primordial temporality and lives out of it. This thesis is treated in the context of the Pauline teaching on eschatology in chapters 4 and 5 of I Thessalonians and in chapter 2 of II Thessalonians.

Heidegger introduces the material with two questions. (1) Granted that Paul and the Thessalonians are bound together in a common religious experience, what is the ultimate factor which possibilizes that experience? (2) Granted that Christian experience is factical life-experience, how is God present (or, to speak phenomenologically, "given") in that factical experience? The two questions receive one answer: Temporality is the modality in which God is "given," i.e., whereby He becomes present, in (intrinsically temporal) factical life-experience. Here is how Heidegger builds the case.

In I Thessalonians 1:10 the characterization of the alreadiness of the Christian reaches a culmination in the sentence: "You have turned to God and away from idols in order to serve a living and true God and to wait for his Son from Heaven." The "turning" and the "serving" receive their meaning from the unifying Christian experience of "awaiting" or "waiting for" the Parousia. The point, then, is to work out the meaning of Parousia and the kind of temporality that goes with it. But before going on, a caution. The importance of the Greek word *parousia* in Heidegger's meditation on Plato and Aristotle is well known. But it is utterly essential, if one is to understand the present passage, that one distinguish the classical Platonic-Aristotelian meaning of *parousia* as a word for beingness as presentness (*ousia*) from the Christian use of the word. There is no confusion in Heidegger, and in what follows we are dealing strictly with the unique *eschatological* meaning which the word has in St. Paul.

First Heidegger sketches the development of the concept of Parousia or Eschaton in the Jewish religion in order to show the unique new meaning it took on in Christianity. He dismisses the idea that scholarship can shed any real light on Christian eschatology by tracing it back through Jewish to Iranian-Babylonian notions of eschatology because, for one thing, such scholarship is already working with a preconceived idea of time (as objectifiably linear) instead of realizing that Christian eschatology revolutionized the concept of time, and, for another, within Christianity itself St. Paul's meaning of Parousia is unique. In this regard Heidegger lays out the development from early Jewish expectation of the "Day of the Lord" through later Judaism's awaiting of the messianic Man to the Synoptic teaching of the coming of the Kingdom of God. In Paul, he claims, Jesus has become the object of eschatological preaching, and hence the Parousia takes on a meaning which cannot be found literally within the word itself. In Paul it means not "presence" or "coming" but *second* coming of the Christ in glory.

But all this is prelude to the real issue, the working out of the meaning of temporality in the light of Pauline eschatology. Stated negatively, Heidegger's basic thesis is the following: The authentic Christian relation to the Parousia is fundamentally not the awaiting of a future event. The structure of Christian hope in the Parousia is radically different from all "awaiting" (*Erwartung*).

Heidegger makes much of the beginning of chapter 5 of I Thessalonians where Paul discusses the question of the "when" of the Second Coming. Paul writes: "Concerning the *chronoi* and the *kairoi* we need not write to you, for you yourselves well know that the day of the Lord will come like a thief in the night..." (5:1-2). The question of the "when" of the Parousia, the *chronos* and *kairos* (Heidegger translates: *Zeit und Augenglick*), is absolutely unique. How is the question answered? Not by any reference to objective time, not even to the time of everydayness.

Rather the question, at it were, is bent back and referred to factical life-experience. "You yourselves well know" (*oidate*)--and this *odiate* refers to that knowledge which the Thessalonians have insofar as they have already become what they are. This is the knowledge inherent in *Gewesensein*, the already-dimension of their existence. The question of temporality in Christian religious experience becomes a matter of how one lives one's facticity. This is not a matter of moralism but fundamentally of temporality.

Paul specifies this further in the next verses.

When people say, "Peace! Security!" then suddenly destruction will come upon them as travail comes upon a woman with child, and there will be no escape. But you are not in darkness, brethren, for that day to surprise you like a thief. For you are all sons of the day; we are not of the night or of darkness. So then let us not sleep...but let us stay awake and be sober.

Here Paul concretizes the question of the temporality of the Parousia by delineating two groups of people. The first are those who urge peace and security. They are, says Heidegger, absorbed in and totally dependent on the world in which they live. Their "waiting" is all for this world. They cannot be saved because *they do not possess themselves,* they have forgotten the authentic self. They live, says St. Paul, in darkness.

The second group lives not in darkness but in the light of day. The word "day," Heidegger claims, has two meanings: first of all, the light of self-comprehension, and secondly, the Day of the Lord itself. The two are bound together. To relate authentically to the Parousia means to be "awake," not primarily to look forward to a future event. The question of the "when" of the Parousia reduces back to the question of the "how" of life--and that is "*wachsam sein*," to be awake. And further, by way of contrast with the first group who cry out "Peace! Security!" the Christian's state of wakefulness in factical experience means a *constant, essential and necessary uncertainty*.

The Christian--or Pauline--meaning of eschatology has shifted from the expectation of a future event to a presence before God, what Heidegger calls a *Vollzugszusammenhang mit Gott*, a context of enacting one's life in uncertainty before the unseen God. The weight has shifted to the "how" of existence. Yes, the Parousia seems to St. Paul to be imminent, but for Heidegger that imminence serves only to characterize the "how" of factical life: its essential uncertainty.

From out of this context of the enactment of factical life-experience before the unseen God (i.e., "being awake") there is generated primordial temporality. The meaning of facticity is temporality, and the meaning of temporality is determined from out of one's basic relation to God. One is becoming, in the uncertainty of the "future," what one has already become. Moreover, Christian religious life is nothing other than the living out of this unique temporality, and, correlatively, only from out of this temporality can the meaning of God (whatever it might be) be determined. And no objectivist conceptions of time can ever touch this kind of temporality. Thus, no Greek philosophical conceptions from Plato or Aristotle can ever be of use here. In fact, Heidegger says that Luther *did* understand this basic experience of temporality and for that reason opposed Aristotelian philosophy so polemically.

Even from this brief sketch we can see that already in 1920–21 the basic lines of Heidegger's doctrine of temporality were set and that they issued not from his reading of the Greeks but from his interpretation of early Christianity. The formal structure is essentially not different from what *Being and Time* will spell out: that the present situation is opened up by two co-equal moments of self-transcendence: becoming oneself understandingly by assuming what one already and essentially is, viz., one's finiteness. More succinctly: the *present* is opened by one's *becoming* what one *already* is. Fundamentally for Heidegger, primordial time is a matter of regaining the "essential" (*Wesen, Gewesenheit*), whether, as in the early Heidegger, that dynamic is described as "coming towards the already-essential" (*Zukommem auf das Gewesene*) or whether, as in the late Heidegger, the dynamic is described as "the arrival of the already-essential" (*Ankunft des Gewesenen*). If we may attempt to capture the same simple phenomenon in another lexicon, we might say that the primordial temporality worked out in *Being and Time* means: bringing to presence one's already operative absence: presence-by-absence, pres-ab-sence. Or in the language of the later period: allowing oneself into (= *Gelassenheit*) the prior absence.

Not only the doctrine of temporality, but other themes as well which will appear in *Being and Time* are incipiently present in this lecture course. *Entschlossenheit* (resolve) has its roots in what Heidegger here calls *wachsam sein* (being awake); *Vorlaufen zum Tode* (anticipation of death) may be seen as taking shape, with significant transformations, out of the *Gegenwärtigen vor*

Gott. But it may be premature and risky to speak of such connections. Between this course and *Being and Time* a lot has yet
to transpire: Heidegger's interpretations of Aristotle in 1921-22,
especially his precision of *Befindlichkeit* (affective disposition)
throughout the interpretation of *pathos* in *Rhetorica* B, and much
more. Above all the being-question, which is scarcely alluded to
here, must be formulated. Yet even if it is premature to make
such connections spanning the years 1921-25, the phenomenon of
"foreshortening," which I mentioned above, also makes them
inevitable.

III

I shall end by returning to the beginning: the letter which
Edmund Husserl wrote to Rudolf Otto in March of 1919 concerning
his book *Das Heilige*. Husserl had some critical words for his
former colleague from Göttingen.

Through Heidegger and Oxner (I no longer know who
took precedence in the matter) I became aware last summer of
your book on the holy, and it has had a strong effect on me
as hardly no other book in years. Allow me to express my
impression in this way: It is a first beginning for a phenomenology of religion, at least with regard to everything that
does not go beyond a pure description and analysis of the
phenomena themselves. To put it succinctly: I cannot share in
the philosophical theorizing which is inserted: moreover, it is
also entirely non-essential for what is proper to this book and
for its specific task, and it would be better left out. It would
seem to me that a great deal more progress must be made in
the study of the phenomena and their eidetic analysis before
a theory of religious consciousness as a philosophical theory
could arise. Above all, one would need to carry out a radical
distinction: between accidental *factum* and the *eidos*. One
would need to study the eidetic necessities and eidetic possibilities of religious consciousness and of its correlate. One
would need a systematic eidetic typification of the levels of
religious data, indeed in their eidetically necessary development. It seems to me that the metaphysician (theologian) in
Herr Otto has carried away on its wings Otto the phenomenologist, and in that regard I think of the image of the angels
who *cover their eyes* with their wings. But be that as it may,
this book will hold an *abiding* place in the history of genuine

philosophy of religion or phenomenology of religion. It is a
beginning and its significance is that it goes back to the
"beginnings," the "origins," and thus, in the most beautiful
sense of the word, is "original." And our age yearns for
nothing so much as that the true origins may finally come to
word, and only then, in the supreme sense, come to their
Word, to the Logos.[14]

Compare the programmatic delineation of a phenomenology of
religion which Husserl gives here with what we have just seen in
Heidegger's course: Husserl's theory of religious consciousness
and its correlates on the one hand, and Heidegger's hermeneutic
of factical life-experience and its correlative worlds on the other;
Husserl's systematic eidetic typification (*Typik*) of the levels of
religious givens on the one hand, and Heidegger's historically enacted situation structured by primordial temporality on the other.
The split between Husserl and his *Lieblingsschüler* was largely
visible in 1920, even though Husserl would not see it clearly until
the summer of 1929, when he finally read *Being and Time* and
scrawled in pencil on the title page: *amicus Plato, magis amica
veritas*, "Plato is a friend, but truth is a greater friend." And
the phrase is, from our perspective, not without some irony.
Husserl seems to speak the phrase *in propria persona* and against
Heidegger, as if it were to read: "Heidegger is my friend, but
truth a greater friend." However, to judge from Heidegger's
1925-26 criticisms of Husserl's separation of the realms of the
ideal and the real, the sentence could easily have been spoken by
the younger man against his master.[15]

Be that as it may, Husserl ended his letter to Otto with the
most beautiful compliment that a phenomenologist could offer: the
work, he says, is a return to *origins*. Apply that to Husserl and
Heidegger. The origins which the master and his understudy
sought turned out to be quite different. For Husserl it was the
transcendental ego as source of meaning-giving for the whole
realm of sense. For Heidegger it was the temporally disclosive
conjunction of man and the primordial absence whence issues the
meaningful presence of all that is. *Ur-ego* vs. *Ereignis*. But
despite the differences which separate the two men, we may
wonder whether Husserl might not have bestowed the same
compliment on his favorite student's effort at a phenomenology of
religion: "It is a beginning and its significance is that it goes
back to the 'beginnings,' the 'origins,' and thus, in the most
beautiful sense of the word, is 'original.'"

NOTES

1. For more information on the early period of Heidegger's development, cf. my essay, "Heidegger's Early Years: Fragments for a Philosophical Biography" in *Listening*, 12 (Fall, 1977), 3-20; reprinted in *Heidegger, the Man and the Thinker*, ed. Thomas Sheehan (Chicago: Precedent, 1979).

2. R II Natorp 7.X.17. I am grateful to Prof. Samuel IJsseling, Director of the Husserl Archives, Leuven, for permission to cite from this and other letters *infra*.

3. R I Natorp 8.X.17. *Ibid.* for the following citations.

4. For an account of the circumstances surrounding Heidegger's nomination to succeed Hartmann at Marburg, cf. my essay, "Time and Being, 1925-27" in *Thinking About Being*, ed. Robert Shahan and J. Mohanty (Norman, Okla., Oklahoma U.P., 1979), pp. 177-219.

5. R I Otto 5.III.19. The original copy of the letter is found in the Rudolf-Otto-Nachlass at the Universitätsbibliothek in Marburg, Hs 797:794. It has been published in an imperfect transcription in Hans-Walter Schütte, *Religion und Christentum in der Theologie Rudolf Ottos* (Berlin: de Gruyter, 1969), pp. 139-142. I have followed the transcription which is to be found in the Husserl Archives in Leuven. For a complete translation, cf. *Heidegger, the Man and the Thinker*.

6. R I Natorp 11.II.20.

7. Announced in the Freiburg Catalogue as *Ausgewählte Probleme der neueren Phänomenologie*, the course had its title changed at the last moment to *Grundprobleme der Phänomenologie*. Heidegger refers to this course in *Sein und Zeit*, p. 72, note 1.

8. R III Walter 20.VI.19.

9. Martin Heidegger, *Logik. Die Frage nach der Wahrheit*, ed. Walter Biemel (Frankfurt: Klostermann, 1976), p. 164.

10. Cf. Martin Heidegger, "Anmerkungen zu Karl Jaspers' *Psychologie der Weltanschauungen*," in his *Wegmarken*, 2nd ed., (Frankfurt: Klostermann, 1976), p. 31. For an excellent commentary on this whole essay (which Heidegger wrote between 1919 and June of 1921), cf. David Farrell Krell, "Towards *Sein und Zeit*," *Journal for the British Society for Phenomenology*, 6 (1975), 147-156.

11. *Wegmarken*, p. 31f. For the previous sentences, *ibid.*, p. 29f.

12. "*Die Situation ist das je in der Entschlossenheit erschlossene Da, als welches das existierende Seiende da ist,*"

Sein und Zeit, p. 299.

13. *"Das Dasein ist...'was' es schon war,"* *Sein und Zeit*, p. 20; *ibid.* for the following sentence.

14. R I Otto 5.III.19.

15. Cf. *Logik. Die Frage nach der Wahrheit*, sections 9 and 10.

4. JACQUES TAMINIAUX

HEIDEGGER AND HUSSERL'S LOGICAL INVESTIGATIONS:
*IN REMEMBRANCE OF HEIDEGGER'S LAST SEMINAR
(ZÄHRINGEN, 1973)**

The introduction to the work that made Heidegger's reputa-
tion and with respect to which he never ceased to gauge the
stages in his path of thinking defines the method of investigation
undertaken in that work as phenomenological. Heidegger recognized his debt to Husserl in the same text when he wrote at the
end of the "Exposition of the Question of the Meaning of Being":
"The following investigation would not have been possible if the
ground had not been prepared by Edmund Husserl, with whose
Logische Untersuchungen phenomenology first emerged."[1] In a
footnote he adds more specifically: "If the following investigation
has taken any steps forward in disclosing 'the things themselves,'
the author must first of all thank E. Husserl, who, by providing
his own incisive personal guidance and by freely turning over his
unpublished investigations, familiarized the author with the most
diverse areas of phenomenological research during his student
years in Freiburg" (SZ 389/489).

The same text, however, combines with this recognition of
debt the expression of a divergence from the "effective turn"
taken by Husserl's investigations and from the "movement" that
claimed him as authority. Consider for example, the following
sentences: "Our comments on the preliminary conception of phenomenology have shown that what is essential in it does not lie in
its *actuality* as a philosophical 'movement.' Higher than actuality

*From Jacques Taminiaux, *Dialectic and Difference. Finitude
in Modern Thought*, edited and translated by Robert Crease and
James T. Decker. Atlantic Highlands, N.J.: Humanities Press
Inc., 1985, pp. 91-114. Reprinted by permission of Humanities
Press International, Inc., Atlantic Highlands, N.J., 07716.

stands *possibility*. We can understand phenomenology only by
seizing upon it as a possibility" (SZ 38, 62f).

Heidegger's investigation alludes often to such a gratitude
devoid of servility and to such an affinity not exempt from
divergence. The most explicit allusion of this kind is found in
section 10, in which Heidegger, distinguishing his Dasein-analysis
from anthropology, psychology, and biology, on the one hand
admits that his plan for such an analysis owes much to Husserl's
teaching, while on the other hand he laments a fundamental omission in Husserl's interpretation of "personality." The Dasein-
analysis is indebted to Husserl for a twofold teaching:

> Edmund Husserl has not only enabled us to understand once
> more the meaning of any genuine philosophical empiricism; he
> has also given us the necessary tools. 'A-priorism' is the
> method of every scientific philosophy which understands it-
> self. There is nothing constructivistic about it. But for this
> very reason *a priori* research requires that the phenomenal
> basis be properly prepared. [SZ 50/490]

The very same analysis diverges, however, from the literal-
ness of Husserl's teaching. Unlike Descartes, who examines "the
cogitare of the *ego*, only within certain limits," Husserl proposes
"to grasp the mode of being which belongs to *cogitationes*." Yet
he fails to see that such a grasp presupposes "the ontological
question of the being of the *sum*," a task that he neglects just as
Descartes does. It is indeed true that Husserl requires for the
unity of the person "a constitution essentially different from that
required for the unity of things of nature," and that he distin-
guishes this constitution from any psychological problematic (since
such a problematic consists in objectifying and in naturalizing the
facts that it treats, and since it fails to recognize that the
Erlebniszusammenhang of consciousness is non-physical in nature
insofar as it is defined by the carrying out of those intentional
acts that are called the *cogitationes*). Nevertheless, the force of
such a requirement and the sharpness of such a distinction are
burdened down by a major weakness and indefiniteness in principle as soon as what is innocently called "the life of the *cogito*" or
the ensemble of "*Erlebnisse*"--no matter how intentional they are
recognized to be--are considered "as a 'given' whose being is not
submitted to questioning" (SZ 46-48/71-73).

Heidegger concedes that Husserl's concrete analyses of the
constitution of "personality" in his *Ideen II* are more rigorous and

employ a more refined conceptual apparatus than those of Dilthey, whom Husserl praises for having "grasped the problems which point the way, and [for having seen] the directions which the work to be done would have to take" (SZ 47-48/72-73); yet, Heidegger equally laments in both cases the fundamental oversight wherein "'life' itself as a mode of being does not become ontologically a problem" (SZ 46/72). Heidegger certainly does not overlook the fundamental difference between the investigations of Dilthey and those of Husserl, namely, the grounding of Husserl's problematic of the personality's constitution on pure intentional consciousness. But this difference remains indecisive; it is indecisive for as long as the mode of being of the *cogito*, as well as that of the "life" by which it is characterized, are understood (in accordance with a long-unexamined tradition) "in the sense of being-present-at-hand (Vorhanden-seins) and of taking-place (*Vorkommens*)" (SZ 48/74; translation modified).

The traces of praise and of divergence are numerous in *Being and Time*, and perhaps the expression of the former takes on the whole a more modest form than that of the latter. Each commendation is pronounced in brief footnotes, whereas each divergence, even when only allusive, is pronounced in the body of the text. Expressions of divergence always aim at the same ontological blindness in Husserl's investigation, at the same lack of radicalness, and at Husserl's constant acceptance of the massive and obscuring privilege of *Vorhandenheit* in his attention to Being.

This divergence is opened up so as to erode such a privilege. Thus, the exposition of Being-in-the-world in general as the constitution of Dasein begins by proposing that "because knowing has been given this priority, our understanding of its ownmost kind of being gets led astray, and accordingly Being-in-the-world must be exhibited even more precisely with regard to knowing the world, and must itself be made visible as an existential 'modality' of being-in" (SZ 60/86). Yet, it is certain that Husserl associated the priority of knowledge with the very idea of phenomenology, and it is precisely the antinaturalistic, antipsychologistic, and presuppositionless style in Husserl's search for an ultimate ground in the immanence of consciousness which is alluded to in these lines:

Now the more unequivocally one maintains that knowing is proximally and really 'inside' and indeed has by no means the same kind of being as beings which are both physical and psychical, the less one presupposes when one believes that

one is making headway in the question of the essence of
knowledge and in the clarification of the relationship between
subject and object. (...) But in any of the numerous varie-
ties which this approach may take, the question of the kind
of being which belongs to this knowing subject is left entirely
unasked, though whenever its knowing gets handled, its way
of being is already included tacitly in one's theme. Of course
we are sometimes assured that we are certainly not to think
of the subject's "inside" and its "inner sphere" as a sort of
"box" or "cabinet." But when one asks for the positive significance of this "inside" of immanence in which knowing is
proximally enclosed, or when one inquires how this "being
inside" which knowing possesses has its own character of
being grounded in the kind of being which belongs to the
subject, then silence reigns. [SZ 60/87]

Heidegger sets allusively a demarcation of the same style
over against the methodical starting point of Husserl's problematic, when he attempts to determine the starting point of the
existential question: "Who is Dasein?" An immediately understandable statement, one that is a matter of course, seems to offer a
sure answer, ready to serve as starting point for the ontological
interpretation of the being that Dasein is. This statement proposes that the "Who" is none other than the "I" itself, the "subject," the "Self." Such a statement is a matter of course, self-
evident. It is an ontical assertion that is understood immediately.
It nevertheless harbors implicitly an unexamined, everyday, and
inherited ontological interpretation concerning the Who in question. That interpretation is therefore not radical and, as such,
diverts questioning from its correct path. According to that
unexamined ontological interpretation--the "I" that, since
Descartes, has been said to be the subject and that to which the
properties of *hupokeimenon* have long since been attributed, itself
having become everyday and just as unexamined--that ego is
thought of as a "given" that is always already there, that persists constantly in a "closed realm," and that "lies at the basis,
in a very special sense, as the *subjectum*" (SZ 114/150).

In short, only that mode of being called *Vorhandenheit* is
attributed to the "I" that is said to be "subject." Access to Heidegger's question, "Who is Dasein?" is consequently barred, since
the question itself implies that *Vorhandenheit* is the mode of being
of that being which is not of the same kind as Dasein. The phenomenological interpretation of Dasein could only diverge from the

path that the question "Who is Dasein?" recommends, in taking as the starting point the "I" thus "given": being-there, present, subsistent, *vorhanden*. It is at this juncture that Heidegger's phenomenological method turns away most deliberately from that "effective reality," the strictures for which Husserl had established. Heidegger is fully aware of these strictures, and in fact he considers and rejects their prejudicial objections in the following lines:

> But is it not contrary to the rules of all sound method to approach a problematic without sticking to what is given as evident in the area of our theme? And what is more indubitable than the givenness [*Gegebenheit*] of the "I"? And does this givenness not tell us that if we aim to work this out primordially, we must disregard everything else that is "given"--not only a "world" that is, but even the being of other "I"s? The kind of "giving" we have here is the mere, formal reflective awareness of the "I"; and perhaps what it gives is indeed evident. This insight even affords access to a phenomenological problematic in its own right, which has in principle the signification of providing a framework as a "formal phenomenology of consciousness." [SZ 115/151]

The effective reality of Husserl's methodical strictures is exactly what is alluded to here. Yet, such strictures, no matter what guarantees of certitude Husserl claims to associate with them, ought themselves to be examined. What if this so-called givenness were to mask rather than to reveal Dasein? And what if Dasein were precisely not a given? Is not being "given" what usually characterizes *Vorhandenheit*? Even if the "I" were to be cut off from its relation to the world and from its ties to others in order to assure its privileged status, it would nonetheless permit the modality of being of *Vorhandenheit* to subsist as eminent, exclusive, and misleading (SZ 115/151).

These considerations and citations suffice to show that the claims of *Vorhandenheit* set the stage for Heidegger's debate with Husserl. Husserl's lack of vigilance with respect to the alleged universal admissibility of the traditionally evident claims of *Vorhandenheit* is what prevents him from posing the question as to the meaning of "Being." That same lack of vigilance also motivates the numerous examples of Heidegger's divergence from the effective turn taken by his teacher's thought.

If so, how is it that the exposition of the question of the

meaning of Being ends with this acknowledgment: "The following investigation would not have been possible if the ground had not been prepared by Edmund Husserl, with whose *Logische Unters-uchungen* phenomenology first emerged"? If Heidegger's reasons for departing from Husserl's teaching in *Being and Time* are evident, it is less evident on the other hand how his elaboration of the question of the meaning of Being can lay claim to a foundation established by Husserl, to a "ground" (*Boden*) that the author of the *Logical Investigations* was to have prepared. On the one hand, when he mentions Husserl's first work in a kind of acknowledgment of affinity, and, on the other hand, when all signs of divergence seem to be aimed at a problematic and a methodology of Cartesian inspiration that Husserl expressly establishes only after the *Logical Investigations*, then Heidegger seems to be claiming that the *Logical Investigations* deserve special attention. Yet, this hardly explains the reasons for such attention.

If, however, we consider those texts to which Heidegger turns several decades later in situating his thinking with respect to Husserl, then we notice that the privileged role of the *Logical Investigations* in the genesis of the question of Being is repeated and confirmed, though the reasons for this priority remain obscure and enigmatic.

Reminiscing in his homage to Hermann Niemeyer in "My Way to Phenomenology," Heidegger insists once again on the importance that the *Logical Investigations* had for the clearing of this way. The journey's start, he insists, was incited by a still imprecise question that arose in 1907 out of his reading of Brentano's dissertation "On the Manifold Meaning of Being in Aristotle" (1862). The first version of the question was, "If a being is predicated in manifold meanings, what then is its leading and fundamental meaning?" Seized by this question, the young student Heidegger, having meanwhile learned that Husserl was in some way a disciple of Brentano, decided already in his first semester of university studies (1909-1910) to seek in the *Logical Investigations* a "decisive aid in the question stimulated by Brentano's dissertation."[2]

It was at first a disappointing wait and search. Heidegger continues, however, "I remained so fascinated by Husserl's work that I read in it again and again in the years to follow without gaining sufficient insight into what fascinated me" (SD 82/75). His captivation in this case was supported by a twofold movement of attraction and reservation: an attraction determined by the feeling that an unprecedented insight was to be found in the book, a reservation determined by a "fundamental difficulty" resulting "from the discordant character (*Zwiespältigkeit*) that Husserl's work presented at first glance" (SD 82/76, translation modified). The book refutes all the pretensions of psychology to ground a theory of knowledge, while at the same time it assigns no less important a place to a kind of psychology, that is, to the description of those essential acts of consciousness from which knowledge is erected.

Heidegger maintains that this indecision in the *Logical Investigations* as to the manner in which phenomenology should be accomplished--that is, as to the relation of the phenomenological task to the realm of consciousness--was removed in 1913 with the publication of [Husserl's] *Ideas*. Here, Heidegger says, "phenomenology consciously and decidedly moved into the tradition of modern philosophy," that is, with subjectivity as the proper site and proper theme of phenomenological inquiry (SD 83f/76). The famous call *Zur Sache selbst*, according to another text of the later period, subordinates the thing [*Sache*] toward which it directs the phenomenologist's attention to the primacy of method, in such a manner that the *Sache* is phenomenologically admissible only to the extent that it satisfies the methodological requirements of validity. These requirements, expressed for the first time by Descartes, assert that the only being whose presence is indisputably valid is the *ego cogito*; in relation to it alone can the being of other beings be determined.[3]

But the text in homage to Niemeyer leaves it to be understood that the *Logical Investigations* were still "philosophically neutral" as to this project of a transcendental phenomenology, which leads to the "reduction" of the Being of beings to an objectivity grounded or constituted in and by absolute subjectivity. The text suggests that it was only later that the *Logical Investigations* found in such a project their systematic setting (SD 84f/76). It was only later that the programmatical explications and methodological clarifications of transcendental phenomenology, consciously and resolutely inscribed by Husserl in the tradition of the modern philosophy of subjectivity, would have imposed the misleading idea that phenomenology has an origin that repudiates all thinking prior to the modern era of subjectivity.

Heidegger does not say that the gap between the *Logical Investigations* and Husserl's later work was something made quickly apparent to him. He recalls rather the quandary, the helplessness, the turmoil, the disquiet that was brought about in him by the question as to what was appropriate to phenomenology and to the manner in which that kind of thinking should be accomplished. His direct contact with Husserl, who in the meantime had taken a post at Freiburg, was going to remove this quandary to a certain extent. But if this contact, in Heidegger's own words, permitted him through practice gradually to learn "phenomenological seeing," it did not weaken at all the sharpness of the attention paid to the *Logical Investigations*, even though Husserl no longer found them perfectly attuned to the systematic setting of the program of transcendental phenomenology. And after becoming Husserl's assistant, Heidegger gave seminars on the *Logical Investigations* for a number of years.

So it is a book that captivated Heidegger for at least fifteen years. Still, Heidegger's published work remains oddly reserved about the why and the wherefore of his persistent fascination for the *Logical Investigations*. The fact that they had a germinative effect is to be accepted if what *Being and Time* called the "ground" upon which the question of the meaning of Being reached its formulation was indeed prepared by this fascination. But this fact does not account for Heidegger's reasons. It may be agreed that his fascination was in no way foreign to the still imprecise question that the reading of Brentano's dissertation had awakened, that his fascination even contributed to the specification of that question. Yet this alone is not enough to see what in Husserl fostered such a refinement. It should be admitted as well that Heidegger repeatedly reexamined the *Logical Investigations*, that his fascination for Husserl's text was accompanied by a simultaneous meditation on Aristotle and the Greeks, which finally resulted in a way of thinking about phenomena far removed from Husserl's deliberately Cartesian position.

Yet, to admit this is to say that Heidegger, at the end of his long consideration of the *Logical Investigations*, had learned that "what occurs for the phenomenology of the acts of consciousness as the self-manifestation of phenomena is thought more originally by Aristotle and in all Greek thinking and existence as *a-letheia*, as the unconcealedness of what is present, its being-revealed, its showing itself." It would be to admit that this attentive meditation on Husserl's text made possible a question that is no longer Husserl's, the question "whence and how is it determined what must be experienced as '*die Sache selbst*' in accordance with the principle of phenomenology? Is it consciousness and its objectivity, or is it the being of beings in its unconcealedness and concealment [*Unverborgenheit und Verbergung*]?" (SD 87/79)

But then, does that imply that his long fascination for the phenomenology of the *Logical Investigations* had been nothing more than the gradual awakening to the fact that a phenomenon is not what Husserl says it to be? Was this a negative fascination, then, providing for the apprentice the occasion to think in opposition to the master? Yet, if that is the case, how can Heidegger's own acknowledgment be understood, that he had discovered the positive "ground" for his investigation in *Being and Time* in the phenomenology of the *Logical Investigations*? Where is the affiliation if the apprenticeship is but the gradual justification for a parting of the ways without return?

All of these questions can be condensed into a single question: Is there something in the *Logical Investigations* that anticipates in a positive sense the *Seinsfrage*, something like the indication of a true break from the obsession for *Vorhandenheit* in Husserl's phenomenology? There is to our knowledge one and only one hint of an answer to this question in Heidegger's published work. In "My Way to Phenomenology," there can be found the following passage:

When I myself began to practice phenomenological seeing, teaching and studying at Husserl's side, experimenting at the same time with a new understanding of Aristotle in seminars, my interest began to be drawn again to the *Logical Investigations*, and especially to the sixth [*Investigation*] in the first edition. The difference between sensuous and categorial intuitions, worked out in that *Investigation*, revealed to me its importance for the determination of the "manifold meaning of Being." (SD 86/78)

This allusion is of capital importance for two reasons. It indicates first of all the site on which Heidegger's fascinated attention finally converged: the sixth *Investigation*, with its distinction between sensuous intuition and categorial intuition. It indicates furthermore that the path that was opened up under Heidegger's fascinated gaze was the very same path leading from the determination of the theme imposed forever upon Heidegger's thinking itinerary by the reading of Brentano's dissertation, the very path for the formulation of the *Seinsfrage*. If Husserl's establishment of the difference between sensuous intuition and categorial intuition had finally proven to be of decisive importance for the formulation of the question of the meaning of Being, then one understands that Heidegger, at a time when Husserl no longer held a very high opinion of his own book, has his "reasons," as he states discretely elsewhere, "to prefer the *Logical Investigations* for the purposes of an introduction to phenomenology."[4] However, nothing in Heidegger's *written* work shows how the difference introduced by the sixth *Investigation* established the "ground" for the *Seinsfrage*. Something, on the other hand, can be found in his *oral* work.

The last seminar held by Heidegger [in 1973, at his home in Zähringen] proposed to attempt to find access to the *Seinsfrage* through Husserl.[5] Its point of departure was a letter [to Heidegger] from Jean Beaufret, who asked in particular: "To what extent may it be said that there is no question of Being for Husserl?"[6] Heidegger developed his answer along two lines. He maintained first of all that, strictly speaking, there is no *Seinsfrage* for Husserl. Because it is a property of the question of Being that it be deployed into the question of the truth of Being, such a question completely eludes metaphysics.

> Metaphysics seeks the Being of beings. Heidegger's question properly aims, if it is legitimate to speak in such a fashion, at the being of Being, or better: it aims at the *truth* of Being (*Wahrheit des Seins*), where *Wahrheit* must be understood according to the safekeeping in which Being is sheltered as Being. In this rigorous sense, there is no *Seinsfrage* for Husserl. Husserl raises strictly metaphysical problems, as for example the problem of the categories. [Z 310f]

Nonetheless, Heidegger concedes at once, "Husserl touches upon, grazes ever so lightly, the question of Being in the sixth chapter of the sixth *Logical Investigation* with the notion of categorial intuition." Particular attention is paid during the first part of the seminar to the task of understanding what it was in this notion that was so "burning" for the formulation of the *Seinsfrage*.

A faithful account of the seminar's proceedings has been published by François Fedier. We would like to attempt in the following pages to confront at our own risk the hints Heidegger provides--hints given in a necessarily allusive manner and rendered accurately in the protocol--with that text from Husserl to which reference is made. In the process, we would also like to examine closely, in the course of the sixth chapter of the sixth *Investigation*, the aspects of that fascination the text exercised over Heidegger at the time he was attempting to articulate his question phenomenologically.

The sixth *Logical Investigation* is entitled: "Elements of a Phenomenological Elucidation of Knowledge." Chapter Six of this *Investigation* begins with a section the general title of which-- "Sensibility and Understanding"--has a Kantian ring, a fact that Heidegger is quick to point out. But the title of the sixth chapter, he adds--"Sensuous and Categorial Intuition"--is not really Kantian at all. For Kant, intuition is strictly bound to sensibility; a categorial intuition would be impossible. A category, in Kant's vocabulary, deduced from the table of judgments inherited from traditional logic, is in no manner given to an intuition. Yet the phrase "categorial intuition" refers to the intuitive givenness of a category. In spite of the Kantian style of the general framework of the sixth *Investigation*, the notion of "categorial intuition" denotes thus a certain surmounting of the Kantian framework, one that presupposes a broadening of Kant's notion of givenness. In order to draw an outline of this anticipation, let us attempt to follow step by step the successive stages of Husserl's interrogation in Chapter Six, to which the seminar referred only allusively.

Generally speaking, the sixth *Investigation* takes as its theme the problem of truth when it attempts to clarify knowledge phenomenologically. In Husserl's terminology, the problem is that of explicating those "objectifying acts" in which the intentionality of consciousness unfolds, insofar as these acts have the function not only of signifying an object emptily, but also of leading to the real manifestation of the object, that is, to what, according to Husserl, the intuitive givenness of the object is in its "identity." The problem of truth is therefore that of the synthesis by which the intuition fulfills the meaning-intention. It is in this context of the problematic of synthesis and fulfillment that Husserl inscribes the sixth chapter.

The question that begins the chapter is formulated in the following way: "What can and must furnish fulfillment for those aspects of meaning which make up propositional form as such, the aspects of 'categorial form' to which, e.g., the copula belongs?"[7] Such a form can be highly diverse in complexity, as is indicated by the utterances and the grammatical forms into which a proposition may be analyzed (substantive, adjective, singular, plural, verb, adverb, etc.); Husserl admits that such utterances and forms refer to distinct instances of meaning. The question is precisely how these instances act in fulfillment. In order to answer this question, Husserl takes the simplest case as provisional basis for examination, as is his custom in the *Logical Investigations*, that of a statement of perception.

Once it has become more specific by this simple basis, the question then becomes, "Are there parts and forms of perception corresponding to all parts and forms of meaning?" If that is the case, answers Husserl, the expression, even if it is composed of a specific "matter"--in other words, the "stuff of meaning"--would be somehow a "mirror," a twin, of the perception. A strict parallelism would reign between perception understood as fulfilling intuition and the specific matter understood as referring to intention. It would of course be conceded that such a parallelism directs the relation of a proper meaning to its corresponding percept. In the case of proper meaning, that is, of the meaning that a proper noun expresses, the meaning-intention corresponds immediately to the percept. "The direct perception" of the city of Cologne "renders the object apparent without the help of further, subordinate acts; it gives the object *which* the meaning-intention means, and *just as* the latter means it. The meaning-intention therefore finds in the mere percept the act which fulfills it with complete adequacy." But in the case of structured and articulated expressions, which therefore are no longer as simplistic as direct and proper naming, such a parallelism is out of the question, according to Husserl.

It is true that one is inclined to believe at first glance that a structured and articulated expression is merely the correctly adequate mirror of a percept: "I *see* white paper and *say* 'white paper,' thereby expressing, with precise adequacy, only what I see" (LU 130/775). This inclination is, however, deceptive. It is misleading, according to Husserl, to believe in a coincidence of expression and perception to the extent that this coincidence leads one to think that meaning resides in perception itself, and instigates the obliteration of the complex chain of those acts inherent to vision, on the one side, and to meaning-intention, on the other. In the case at hand, several acts of meaning-intention are linked: that intention directed to the white of the paper, that directed to the paper as substance, and that directed to it as a being. Yet, these intentions, rendered by those words used to express them, extend beyond the contents provided by pure and simple perception. As such, the contents are a rush of pure sensory givens--a *Stoff*, as Husserl calls it in the *Logical Investigations*, or a *hule*, in the terminology of the later works. Take, for example, the intention expressed by the word "white": "The intention of the word 'white' only partially coincides with the color-aspect of the apparent object; a surplus of meaning remains

over, a form which finds nothing in the appearance itself to confirm it." There is an excessiveness of the white meant by the
expression, with respect to the intuitively given moment of color.

There is likewise an excessiveness of what is signified by
the substantive term "paper," with respect to the sensory given.
Husserl calls this excessiveness a "form." Meaning functions as a
form which exceeds the content of pure perception. Far from
expressing a pure and simple seeing, a sentence thus expresses a
meaning that extends beyond what is simply seen. At the same
time, it is true that acts of relating, of unification, of formation
are carried out on the basis of pure and simple perception, that
is, the pure reception of the sensory givens. Such acts, based
upon the reception of the given, have a function of formal
grounding with respect to the given, and by virtue of this function they permit the perception to have a cognitive essence; they
act in such a way that in perception, "the apparent object
announces itself as self-given" (LU 131/776). The meaning-
intentions denoted by the variable forms of expression are to be
compared with these formative, objectifying acts, and not at all
with the pure and simple sensory given. The parallelism of the
naive theory, which sees in the expression the simple mirror of
what is intuitively given, finds itself therefore not so much
repudiated as displaced. The parallelism no longer obtains
"between the meaning-intention of expressions and the mere percepts which correspond to them: it is a parallelism between
meaning-intentions and the above-mentioned *perceptually founded
acts*" (LU 131/776). Yet, at the same time that it finds itself
displaced, that is, reestablished at another level, the parallelism
is just as profoundly transformed. It no longer has the sense of a
simple mirror, since in it that which mirrors exceeds that which
is being mirrored, that is to say, since the meaning-intention is a
kind of excess.

Nevertheless, this strange parallelism, in which the founded
is in turn founder and excessive with respect to that upon which
it is founded, in no way occasions in Husserl's investigations a
questioning of the perspective that governs it. Husserl's perspective is and remains that of the synthesis of fulfillment between a
meaning-intention and an intuition. Knowledge as such has for
Husserl the character of fulfillment and identification; and since
sensuous intuition, even in the case of simple statements of perception, cannot exercise the role of fulfillment, his problem is to
extricate a nonsensuous intuitive field that might exercise this
role.

Such being the general perspective of the problematic and of its solution, it is first of all a matter of defining more specifically what--in the case of statements of perception--the "matter for perception" is and what the "matter for meaning" is. At the same time, we need "to become aware that, *in the mere form of a judgment, only certain antecedently specifiable parts of our statement can have something which corresponds to them in intuition, while to other parts of the statement nothing intuitive can possibly correspond*" (LU 135/778; emphasis is Husserl's).

Perceptual statements articulate utterances of varying pattern, "such as 'A is P,' 'An S is P,' 'This S is P,' 'All S are P,' etc." In such forms of judgment, it is easy to see, Husserl says, "that *only at the places indicated by letters (variables)* in such 'forms of judgment,' can meanings be put that are themselves fulfilled in perception, whereas it is hopeless, even misguided, to look directly in perception for what could give fulfillment to our supplementary formal meanings." Yet, in fact, "the same difference between 'matter' and 'form'" repeats itself even in letters (variables), as can be seen when they are replaced by the unitary terms for which they are the substitutes. Once the variable (subject or predicate) is replaced by a term (this or that substantive, this or that adjective), this term embraces simultaneously those material elements that find their direct fulfillment in sensuous intuition and those "supplementary forms" that do not find direct fulfillment. Glancing over what Husserl calls "the whole sphere of objectifying presentation," there is thus a fundamental difference between the sensuous matter of representation, the *Stoff* or *hule*, and the categorial form of representation. In other words, the objective correlates of categorial forms are not "real" (*real*) moments given as matter with the same sensuous status as the perceived given (LU 135f/778f).

The same applies in a particularly clear manner to the category of "Being," "the form-giving flexion," whether understood in an existential sense or in an attributive and predicative sense. Kant had always insisted that "being is no real predicate." Husserl recalls and agrees with this thesis, as it aims precisely at what Husserl himself seeks to clarify. And he gives the following commentary to that thesis:

I can see color, but not *being*-colored. I can feel smoothness, but not *being*-smooth....Being is nothing *in* the object, no part of it, no moment tenanting to it, no quality or intensity of it, no figure of it or no internal form whatsoever, no constitutive feature of it however conceived. But being is also nothing attaching *to* an object: as it is no real internal feature, so also it is no real external feature, and therefore not, in the *real* sense, a 'feature' at all. For it has nothing to do with the *real* forms of unity which bind objects into more comprehensive objects, tones into harmonies, things into more comprehensive things or arrangements of things (gardens, streets, the external phenomenal world). On these real forms of unity the external features of objects, the right and the left, the high and the low, the loud and the soft, etc., are founded. Among these anything like an 'is' is naturally not to be found....For all these are perceptible, and they exhaust the range of possible percepts (*Wahrnehmbarkeiten*), so that we are at once saying and maintaining *that being is absolutely imperceptible*. [LU 137f/780f]

To be imperceptible means that Being is not something that is the "given" of sense-perception in general, neither of external nor of internal perception.

  Being is the given correlate neither of an external sense-perception--we do not see it with our eyes, touch it with our hands, hear it with our ears, etc.--nor for that matter of an internal sense-perception. It is not internally lived, it is not an *Erlebnis*. Husserl insists emphatically that, contrary to Locke and the empiricist tradition, the origin of the concept "Being," like that of the other categories (the "a" and the "the"; the "and" and the "or"; the "if" and the "then"; the "all" and the "none"; the "something" and the "nothing," etc.), no more resides in the realm of internal perception than in that of external perception. If reflection upon psychically lived experiences yields concepts such as affirmation, denial, collecting, counting, etc., if these concepts are for this reason sensuous concepts, the same does not apply to categorial concepts, which are neither concepts of lived mental acts nor of their real constituents. "The thought of a real judgment fulfills itself in the inner intuition of an actual judgment, but the thought of an 'is' does not fulfill itself in this manner. Being is not a judgment nor a constituent of a judgment" (LU 139/782). The little word "is" certainly is a real constituent of judgment, but the "'is' itself" is completely different from this real constituent. It is in no way a part of the judgment; "the 'is' itself does not enter into the judgment, it is merely meant, significatively referred to, by the little word 'is.'" (LU 140/782)

  The "'is' itself" is nonetheless given, even if not as a real constituent of the given correlate of external sense-perception, nor as a real constituent of internal lived experience, and in particular of judgment as lived mental experience. There is givenness of the "'is' itself" when a state of affairs, one solely envisioned in the meaning-intention inherent to the judgmental act, fulfills the judgment, makes it true or adequate. "Not only what is meant in the partial meaning *gold*, nor only what is meant in the partial meaning *yellow*, itself appears before us, but also *gold-being-yellow* thus appears" (LU 140/782). Such a state of affairs is not in the judgment itself; rather, it is its objective correlate. The relation of this state of affairs, through which the "'is' itself" is given to the act of *Gewahrwerdung* or of becoming aware to which it is adjusted, is conceived by Husserl according to the model of the relation of the sensuous given to sensuous intuition. But the analogy leaves intact the idea of excessiveness precisely obtained from the case of the statement about perception. The concept "red" exceeds any particular sensorially given red. It is the sensuous presence of any particular red that grounds the act of abstraction that permits the realization of the concept, but this concept is inversely that by virtue of which such a particular red is recognized as red. The particular red is given only because red as such, the concept "red," is itself given, if only in a givenness that itself is not sensuous. The same holds for every concept, especially the concept "Being." "It is in fact obvious from the start," says Husserl, "that, just as any other concept (or Ideas, Specific Unity) can only 'arise,' i.e., become *self-given* to us, if based on an act which at least sets some individual instance of it imaginately before our eyes, so that the concept of Being can arise only when *some being, actual or imaginary, is set before our eyes*" (LU 141/784).

The distinction between sensuous intuition and categorial intuition appears at this juncture in Husserl's analysis. The distinction itself is Husserl's answer to the question of "where do the categorial forms of our meanings find their fulfillment, if not in the 'perception' or 'intuition' which we tried provisionally to delimit in talking of 'sensibility'" (LU 142/784), that is, in "perception" or "intuition" in the sense of an affection by a given sensuous matter, or a sensorial *hule*. Husserl's thesis, in response to this question, is that the meanings of categorial forms really find their fulfillment, not, of course, in perception understood in the strict sense of the reception of sensorial *hule*, but in "an act which renders identical services to the categorial elements of meaning that merely sensuous perception [the reception of *hule*] renders to the material elements [the *hule* itself]" (LU 142/785). Categorial intuition is this act, and the recognition of this categorial intuition corrects the theory of "simple perception" from which the analysis began. For it is important to notice that such an act of categorial intuition is co-present simultaneously with sensuous intuition, in the limited sense of the term, whenever a perception takes place--that is, a perception not in the strict sense but in the sense of a "fulfilling act of confirmatory self-presentation" (LU 142/785). Husserl pursues the same line of argument when he writes: "We have taken it for granted that forms, too, can be genuinely fulfilled, or that the same applies to variously structured total meanings, and not merely to the 'material' elements of such meanings, and our assumption is put beyond doubt by looking at each case of faithful perceptual assertion" (LU 142/784). In other words, the object itself is brought before us, is perceived, precisely in forms, in categorial structures. Perception, no longer understood in the strict sense as a reception of a hyletic given, but rather in its cognitive entirety as an act by which the thing itself is identified and con-firmed in its identity, joins sensuous intuition to categorial intuition.

It seemed useful to us to restore at length in its successive stages the interrogation that Husserl conducts in the sixth chap-ter, because Heidegger's commentary on Husserl's theory of cate-gorial intuition, far from being sufficiently supported by the new citations to which he drew the attention of the seminar's partici-pants, is not fully intelligible until it is related to the movement of the chapter as a whole. What first of all is so surprising in Heidegger's commentary is his discretion, which, in the dialogical movement of the seminar, seemed to us to indicate, on the one hand, a kind of modesty in expressing the extent of insight that this fascinating text had exerted on him and, on the other hand, an invitation to the participants, armed with only a few signs, to reconstruct on their own the path that led from Husserl's text to the *Seinsfrage*. It is in this second sense, at any rate, that we would like to consider Heidegger's remarks.

These remarks gravitate around the notion of categorial intuition, which, according to Heidegger, "graze ever so lightly" the *Seinsfrage* and which constitute the "burning issue" in Husserl's thinking. Heidegger develops his commentary via two questions. The first question interrogates the path that led Husserl to this notion and to the victory over the tradition that the notion represents. The answer is suggested by the setting in which Chapter Six (the second section of the sixth *Investigation* on "Sensibility and Understanding") is inscribed, and by the title of the chapter itself, ("Sensuous and Categorial Intuitions"): Husserl sets out from sensuous intuition in order to arrive at categorial intuition. It is, therefore, first of all necessary to find out what sensuous intuition is for Husserl.

Heidegger insists that sensuous intuition is strictly speaking not the perception of a thing, but rather the perception of sensory givens, the affection by the *hule* and its givens (blue, black, extension, etc.). The thing, that is, the "object" of sense-perception from Husserl's perspective, is not given in the *hule*. It is not part of the *hule*. And yet, the object is after all perceived. As Husserl notes, "With the sensory givens in perception comes the appearance of an object"--that is, according to philosophical tradition, of a "substance." But, according to the same tradition, substance is a category--as, for example, in Kant. Yet, contrary to Kant, for whom the category "substance" is a mere form that organizes the diversity of sensibility for the understanding and, as such, is not itself given, Heidegger insists that Husserl thinks of the categorial as a given: the term "categorial intuition" itself suggests a being-present to the category.

Let us refer to the text of the protocol: "I see this book. But where in the book is its substance? I do not see it at all as I see the book. And yet, the book is a substance pure and simple, one that I must in some way see, without which I could see nothing at all" (Z 314). Here we arrive at Husserl's notion of *Ueberschuss*, excess. Heidegger explains: "the 'is'--by which I note the presence of the ink-well as object or substance--is 'in excess' among the sensuous affections: the 'is' is not added to the sense-impressions, but is 'seen,' even if it is seen in a way other than that of what is visible. In order to be seen thus, it is necessary that it be given" (Z 313). This commentary is discreet. At the same time that it brings out the burning character of Husserl's notion of an excessiveness of Being, it blurs to a certain extent the same excessiveness by associating it with the excessiveness of the category of substance. Moreover, the commentary continues to treat the category of substance. "Let us repeat once again: when I see this book, I do indeed see a substantial thing, without seeing however the substantiality as I see the book. Yet it is the substantiality in its non-appearance which allows what appears to appear. In this sense, it can be said that substantiality is more apparent than the apparent itself" (Z 314).

We shall soon attempt to clarify the reasons for Heidegger's discretion. At any rate, the first stage of Heidegger's commentary consists in showing that the path by which Husserl arrives at the notion of categorial intuition is that of *analogy*. There is a categorial intuition analogous to sensuous intuition. The victory that this notion represents over the tradition lies in the explosion of the framework in which the tradition--specifically, the tradition that claimed Kant as its authority--limited the given. The given, according to this tradition, is that which affects the sensibility; categories become for Husserl in turn "as encounterable as anything given to the senses" (Z 314).

Heidegger comments: "This is Husserl's decisive discovery." And it is here that the commentary's second question arises: in what sense was the discovery an "essential stimulus" for the formulation of the *Seinsfrage*? In debating this question, Heidegger emphasizes that the decisive character of Husserl's discovery, far from residing in the contribution of an exhaustive answer to the question already prepared, consists, on the contrary, in the elucidation of an "essential difficulty." Heidegger comments elliptically that this "difficulty stems from the two-fold meaning of the word *to see*....The difficulty consists in the fact that if I see some white paper, I do not see the substance 'as' I see white paper" (Z 314). This, too, is a discreet commentary, for Heidegger's question is less that of the excessiveness of substance with respect to the sensory givens than that of the excessiveness of Being over that which is. The latter excessiveness--which is not dissociable from the first, however, in the sense that it grounds the experience that recognizes the presence of the thing perceived as substance--the excessiveness of Being over that which is, is recognized by Husserl, as we have seen, and it is this excessiveness that motivated Heidegger's fascination. Heidegger continues:

In his analyses of categorial intuition, Husserl freed Being from its fixity in judgment. In so doing, the entire field of inquiry finds itself reoriented. If I pose the question of the meaning of Being, already I must be beyond Being understood as the Being of beings....Husserl's *tour de force* consisted precisely in this presencing of Being [inasmuch as it is beyond the Being of beings] made phenomenally present in the category. By means of this *tour de force*, I was finally in possession of a ground: *Being* is no mere concept, a pure abstraction obtained thanks to the work of deduction. [Z 315]

"I was finally in possession of a ground," that is, a *Boden*, the same word Heidegger used in *Being and Time* when he asserted that his investigation would not have been possible without the foundation (*Boden*) established by Husserl in the *Logical Investigations*.

If one admits that Heidegger's discretion is proportional to the importance of those perspectives made possible by the long fascination that the sixth *Investigation* elicited, Heidegger's commentary seems to invite a consideration of the *Seinsfrage*'s formulation as a kind of retrieval, in the sense in which Heidegger himself uses the term, of Husserl's doctrine of categorial intuition. It likewise invites the recognition that the famous distinction Heidegger draws between the possibility and the actuality of phenomenology is already in Husserl's text.

Let us recall the structure of the *Seinsfrage*. The *Gefragtes*, that which is posed by the question, that about which the question asks, that which is asked in it, is the Being of beings, insofar as this Being is not itself a being. That toward which the question is aimed, the question's intention, its *Erfragtes*, is the meaning of Being, the meaning that *Being and Time* was to find in temporality. That in view of which the questioning is presented, that to which the questioning is addressed, its *Befragtes*, is the totality of beings. Yet the investigation initially focuses, not upon any being whatsoever, but rather first and foremost upon the being that stands in relationship to the question, the being whose mode of being is to understand Being: Dasein. Because the comprehension of Being is the fundamental determination of its being, such a being is not merely given as one being among others. The analysis of Dasein will show that its mode of being consists in being beyond what is merely given; it is a radically ec-static mode of being that makes of Dasein the volatile site of finite transcendence, of this thrown project where the understanding of Being is tied to a dependence upon beings, and where the movement by which the being [Dasein] is made to transcend toward Being is itself solicited by Being, by the openness at whose core that being can manifest itself as a being.

This structure of the question of Being together with certain essential elements in the analysis initiated by that question are, in a certain sense, anticipated in Husserl's doctrine of categorial intuition.

The main assertion of this doctrine concerns the category's excessiveness compared with what is merely given. Yet, not everything is at the same level in this categorial excessiveness.

The category "substance" is not on the same level as the category "Being." It is this latter that, in fact, grounds the objectifying function of the category "substance." When Heidegger says discretely that "substantiality is what, in its non-appearance, allows what appears to appear," it may be added justifiably, in strict agreement with Husserl's text, that it is by virtue of the category "Being" that a state of affairs can prove to *be* substantial and can appear as such. Beings can appear as they are, can let themselves be seen, can display their identity or their truth, by means of an excessiveness of Being over what is merely given, that is, by means of a precession of Being, of an a priori of Being. But Being understood in its a priori excessiveness is no less given than what is merely given, even if it is given otherwise. No matter how excessive Being may be, it is nonetheless given to an intuition. The category "Being" is therefore apparent, it is a phenomenon, and because its function is foundational with respect to what appears, it can be said "that it is more apparent than the apparent itself," that it is the phenomenon of all phenomena, the phenomenon par excellence of phenomenology, that toward which what will soon be called the "reduction" must be led. There is to be found here something like a duplicity of the phenomenon, one strictly interdependent with a duplicity of seeing.

The excess of Being over the given, far from preventing Being from being as much a phenomenon as the given, bestows upon Being the privileged status of originary phenomenon. Things appear as being what they are by virtue of Being. Correlatively, there is an excessiveness at the very heart of seeing. The seeing that grasps things in their coming-into-appearance, intentionality, is itself transfixed by excessiveness; seeing must be beyond the given in order for things to be and to be what they are. And this movement of surmounting is interdependent with an exposure to beings, since, as Husserl writes, "the concept 'Being' can arise only if some being is placed before our eyes." What emerges from Husserl's analysis is that the correlation of the duplicity of phenomena with the duplicity of seeing frees Being from its fixity in judgment, and that this correlation constitutes that toward which knowledge knows, something that transcends both its acts and their correlates.

Heidegger's long-fascinated gaze found in the *Logical Investigations* the emergence of a group of themes that animate the *Seinsfrage* in *Being and Time*: namely, that Being transcends beings; that Being is the *transcendens* par excellence; that Being is in a privileged sense *the* phenomenon of phenomenology; that the coming-into-appearance of beings requires a prior understanding of Being; that this very understanding, to the extent to which it is always beyond, is nonetheless inseparable from an exposure to beings; and that the excessiveness of Being is the cradle of truth. Above all, it is in the excellent way by which Husserl's analyses approach the question of Being that they earn Heidegger's praise for having given to philosophy its authenticity, "empirical apriorism."

There exists, therefore, a properly Husserlian version of the ontological difference. If this difference is accounted for in the theoretical setting in which it occurs for Husserl, it can be said that it consists in grounding upon the excessiveness of Being the identification of the *Sachverhalt*, understood in its very presence. It consists as well in furnishing at the same time an excessiveness within the compass of that intentionality for which the identification takes place. The difference in consciousness between the real contents of psychical life, the *Erlebnisse*, insofar as they are merely given, and the objectifying acts, given in a completely different way, exceeding the *Erlebnisse* and conferring upon them an intentional status, corresponds to the difference between Being--which, says Husserl, is "absolutely imperceptible" and yet "intuited"--and the given.

But the whole difficulty--and it is enormous, which explains Heidegger's almost compulsive fascination for it--is precisely that of knowing whether these concepts and the theoretical setting that organizes them are sufficient to account for the excessiveness in question, and whether this setting does not inevitably cover up once more what Husserl discovers. If it is admitted that the a priori excessiveness of Being is originary, which is what Husserl seems to anticipate when he suggests that no state of affairs could be recognized for what it is without such an excessiveness, then should not the traditional language of the philosophy of consciousness, together with all the conceptual pairs that continue to circulate in Husserl's text--subject-object, immanence-transcendence, form-content, emptiness-fulfillment, activity-passivity--be shattered? How could this language still be considered pertinent if it is true that Husserl, whatever pains he may take to distinguish the psychological from the ideal--that is, from what he will later call the transcendental--cannot avoid assigning these pairs to the very same ontical order, abolishing in the process the originary excessiveness? Does not Husserl's thesis--that a category is given otherwise than as the form of a sensuous content--lead to the leveling of the excessiveness into the sole order of *Vorhandenheit*, merely because the mode of being of the category is traced analogically from the mode of being of sensory givens? Does that not likewise imply that Being has no other meaning than that of being an object?

The intrinsically relational structure of what Husserl calls intentionality is in a sense what seems to refuse such a leveling. But if, as is the case in the setting for Husserl's retrieval of the Cartesian *cogito sum*, intentionality is attributed to a subject that is in turn grasped as a given, as a *Vorhandenes*, then is not the excessiveness once again abolished? Can intentionality be attributed to a subject, and is the strictly ontical division of interiority and exteriority still appropriate, if intentionality is essentially a relation to the Being beyond beings, a relation that transfixes the very perception of beings? If our relation to a being's manifestation presupposes that Being itself is a phenomenon, is it not then necessary to find names other than that of consciousness, which in itself designates nothing other than the interiorization of *Vorhandenheit*, the re-presentation of the given, in order to designate the locus where this relation transfixed with excessiveness emerges? And why must the study of this relation become fixed in the narrow confines of a theory of knowledge, if it is really originary for the being that we are, if therefore it permeates all our dealings with things? And can this relation to the manifestation, as the disclosure of beings that presupposes the a priori and excessive truth of Being, still be described in the ontical metaphors of emptiness and fulfillment? And is not the synthesis of fulfillment, according to which Husserl interprets the givenness of a state of affairs itself in its identity, understood as the adjoining of one being to another being, covering up at the same time the excessiveness of Being?

It is not our intention to follow these questions, which are far from being the only ones implied in Heidegger's fascination for the sixth *Investigation* and which occasion numerous divergences from Husserl in *Being and Time* as well as in the Marburg courses on phenomenology.[8] Our purpose was solely to indicate in which direction these same divergences could still claim a "ground established by Husserl."[9]

One final word to conclude and to correct the partiality in what has just been said. From this ground to these same divergences, the relationship of Heidegger and Husserl is in no way that of a simple inference. It is never a question for Heidegger, as our remarks might lead one to believe, of taking one of

Husserl's theses as exemplary so as to extract from it implications
by which to arrive at the unacceptability of other theses. If it is
tempting for a historian of philosophy to construct such parallel
logical schemes, nonetheless such schemes in no way encompass
what occurs when one thinker's interrogation awakens that of
another thinker. When Heidegger sees the prospects of the
Seinsfrage being outlined in Husserl's sixth *Investigation*, so
often reread, it is an issue of something entirely different from
the birth of a thesis from which other theses, especially Husserl's
theses, might be criticized. It is, instead, an issue of the birth
of a question that will have no end, with which he will never
cease to be associated, and that owes its force to the fact that it
remains a question.

In the movement of this question, ceaselessly retrieved,
ceaselessly displaced, it was necessary to show that the "ground"
that made *Being and Time* possible--thanks to Husserl and extending beyond him--was nothing upon which something could be
grounded, and that the project of a scientific philosophy that
understands itself, in the form of a fundamental ontology, barred
access to a proper attention to the question of Being. The
"ground" was therefore no ground at all. Or rather: it was indeed such a ground. For why should "ground" mean foundation?
Does not grounding imply setting one being upon another? Whereas what was this "ground" if not the excess of Being over beings?

At the end of this movement, and from the famous *Kehre* on,
Husserl is hardly mentioned. Yet, this silence--another aspect of
Heidegger's discretion--does not prevent the old fascination from
continuing to vibrate. The following passage from one of Heidegger's last great texts--"Time and Being"--will be familiar to those
who recall the sixth *Logical Investigation*:

> *Is* Being at all? If it were, then incontestably we would have
> to recognize it as something which is (*als etwas Seiendes*) and
> consequently discover it as such among other beings. This
> lecture hall *is*. The lecture hall *is* illuminated. We recognize
> the illuminated lecture hall at once and with no reservation as
> something that is. Yet where in the whole lecture hall do we
> find the 'is'? Nowhere among things do we find Being. [SD
> 3/3]

And when Heidegger closes his last seminar with a reinterro-
gation of the famous *esti gar einai* in Parmenides' poem, of this

Sachverhalt that ought to be heard with a Greek ear and thus understood as *anwesend-anwesen*, he ended his reading with the following words: "This thought of Parmenides is grounded upon that which has appeared to seeing. As Goethe notes, the greatest difficulty perhaps is to achieve a pure remark [*reine Bemerkung*]. It is precisely this difficulty which is at stake for Parmenides: to come to take into view *das anwesend-anwesen*.... The thinking demanded here I call tautological thinking. This is the originary meaning of phenomenology" (Z 336ff). Those words--*Sachverhalt*, pure remark, seeing, taking into view, identity--are the very fabric of the *Logical Investigations*. It is now and always the old fascination before the interdependence between the excessiveness of seeing and the excessiveness of the phenomenon that incites the final session.

NOTES

1. Martin Heidegger, *Sein und Zeit*, (Tübingen: Max Niemeyer Verlag, 8th ed., 1957), p. 38; *Being and Time*, (New York: Harper & Row, 1962), p. 62. Subsequent references will follow in the text as SZ, followed by the German and English pagination. Thus:(SZ 38,62).

2. *Zur Sache des Denkens*, (Tübingen: Max Niemeyer Verlag, 1969), p. 82, "My way to Phenomenology," in *On Time and Being*, trans. J. Stambaugh (New York: Harper & Row, 1972), p. 74f. Subsequent references in the text as SD followed by the German and English pagination.

3. "*Das Ende der Philosophie und die Aufgabe des Denkens*," SD, 69-71(62-65).

4. *Unterwegs zur Sprache* (Pfüllingen: Neske, 1959), pp. 90f; "A Dialogue on Language," in *On The Way to Language*, trans. P. Hertz (New York: Harper & Row, 1971), p. 5.

5. [Protocols for each of the three Zähringen sessions were prepared by the participants, who included, in addition to Heidegger, Beaufret, and Taminiaux, French philosophers François Fédier, François Vezin, and Henri-Xavier Mongis. The text available in *Questions IV* consists of reworked protocols of all the three sessions prepared by Fedier in French; the German edition of the seminar, included in *Vier Seminare*, ed. C. Ochwadt (Frankfurt am Main: V. Klostermann, 1977), is in fact a German *translation* of the Fédier protocols and thus cannot claim to be closer to the original events of the sessions than the French

account. This situation presents obvious interpretive difficulties. As a participant able to rely upon the combined resources of his own notes, the earlier protocols, and the Fédier text, Taminiaux might be expected to be less vulnerable to those difficulties. Because no translation of the Zähringen text is yet available in English, we have supplied our own translations based on the French text. Subsequent references to the seminar will be cited in parentheses as follows: (Z 309)--eds.]

6. "Le séminaire de Zähringen, in Questions IV (Paris: Gallimard, 1976), pp. 309f. Note, however, this indication from the protocol of a seminar given in conjunction with Heidegger's lecture "Time and Being": "Husserl himself, who came close to the true question of Being in the Logical Investigations--above all in the VIth--could not persevere in the philosophical atmosphere of that time." From "Summary of a Seminar on the Lecture 'Time and Being'", in Time and Being (New York: Harper & Row, 1972), p. 42.

7. Edmund Husserl, Logische Untersuchungen, 4th ed. v.II, (Tübingen: Max Niemeyer Verlag, 1968) part 2, p. 129; Logical Investigations, trans. J. N. Findlay (New York: Humanities Press, 1970), v.II, p. 774. Subsequent references will be cited in the text as LU, followed by the German and English pagination.

8. See the Marburg lectures, Winter Semester 1925/26, in Gesamtausgabe Band 21, entitled Logik, Die Frage nach der Wahrheit, in particular the preliminary remarks concerning psychologism and the question of truth, and the questions concerning Husserl's concept of truth in the sense of identity (Section 10). See also the course for the summer semester of 1927, Die Grundprobleme der Phänomenologie, Gesamtausgabe Band 24, which contains notably a "phenomenologico-critical discussion" of Kant's thesis, which, as we have seen, inspired Husserl in his sixth Investigation: that Being is not a real predicate. It would be easy to read this "discussion"--during which the phrase "ontological difference" appears probably for the first time--as a "retrieval" of the sixth Logical Investigation, by means of a debate with Kant.

9. For a general view of the relationship between Heidegger and Husserl, see the excellent study by Jean Beaufret, "Husserl et Heidegger" in Dialogues avec Heidegger, vol. 3 (Paris: Ed. de Minuit, 1974), pp. 108-154.

5. JOHN SALLIS

THE ORIGINS OF HEIDEGGER'S THOUGHT*

> Delphi schlummert und wo tönet das
> grosse Geschik?
> Friedrich Hölderlin
> "Brod und Wein"

Heidegger was among those for whom the untimely death of Max Scheler in 1928 brought an experience of utter and profound loss. In a memorial address, delivered two days after Scheler's death, Heidegger paid tribute to Scheler as having been the strongest philosophical force in all of Europe and expressed deep sorrow over the fact that Scheler had died tragically in the very midst of his work, or, rather, at a time of new beginnings from which a genuine fulfillment of his work could have come. Heidegger concluded the address with these words:

> Max Scheler has died. Before his destiny we bow our heads; again a path of philosophy fades away, back into darkness.[1]

Heidegger's death, however, seems different. It came not in the midst of his career but only after that career had of itself come to its conclusion. His last years were devoted to planning the complete edition of his writings, and he lived to see the first two volumes of this edition appear. The reception of his work seems likewise to have run its course, from violent criticism and

---

*Text of a memorial lecture presented at the University of Toronto on October 21, 1976 and at Grinnell College on November 12, 1977. The text was published first in *Research in Phenomenology*, 7(1977), pp. 43-57. Reprinted by permission of Humanities Press International, Inc., Atlantic Highlands, N.J. 07716.

misunderstanding to an appreciative assimilation of his work. Today Heidegger's thought is acknowledged as having been a major
intellectual force throughout most of this century--a force which
has drastically altered the philosophical shape of things and given
radically new impetus and direction to fields as diverse as psychology, theology, and literary criticism. But now, it seems, that
impact is played out. Heidegger's thought, now assimilated, is being enshrined in the history of philosophy. It is as though a
well-ripened fruit had finally dropped gently to the ground.

Perhaps, however, the death of a great thinker is never
totally lacking in tragedy. For even if his life is lived out to its
conclusion, as was Heidegger's, his work is never finally rounded
out. The case of Socrates is paradigmatic, the philosopher engaged in questioning even throughout his final hours, exposing
himself to the weight of the questions asked by his friends, and,
most significantly, letting his positive thought, his "position," be
decisively fragmented by a great myth just as it is about to be
sealed forever. The work of a genuine thinker never escapes the
fragmentation, the negativity, to which radical questioning
exposes him; and death, when it comes, seals the fragmentation
of his work. Death fixes forever the lack, the negativity, and
testifies thus to the inevitable loss by casting that loss utterly
beyond hope. Death brings philosophy to an end without being its
end, its telos, its fulfilling completion. Death stands as a tragic
symbol.

Even in its external appearance Heidegger's work is fragmentary. His book *Being and Time*, first published in 1927, remains unfinished, even though he always considered it his
magnum opus. Moreover, its unfinished state is precisely such as
to leave the entire project in suspension, for that part which remains unpublished is just the one in which Heidegger would actually have carried through the task for which the entire published
part is only prepatory. Heidegger did not succeed in determining
the meaning of Being in terms of time, and by ordinary standards
Being and Time, falling short of its expressed goal, is a failure--
or, at best, a torso. That failure is not redeemed by the later
writings. To the extent that they take up and deepen the questioning of *Being and Time*, they do so only at the expense of still
greater fragmentation. With Heidegger's death too a path of philosophy fades away, back into darkness. Though in a different
way, Heidegger's work remains as tragically fragmentary as
Scheler's

Our response to Heidegger's death can be thoughtful--rather

than merely biographical--only if we re-enact, as it were, a strand of this tragedy. To this end we need to release Heidegger's work from that seal of fragmentation brought by his death; that is, we need to let that fragmentation assume the positive aspect which it has in living thought. What is this positive aspect? It is that aspect which Heidegger designated by referring to his thought as *under way*.

If we would re-enact such thought, it is imperative that we understand what set it on its way--that in response we might set out correspondingly. It is imperative also that we understand what sustained it on that way, what shaped the way itself--that we too might keep to that way. We need, in other words, to understand the origins of Heidegger's thought.

This, then, is our primary question: What are the origins of Heidegger's thought? We shall deal with this question at three progressively more fundamental levels. These three levels correspond to three distinct concepts of origin. Initially, we shall take origin to mean historical origin and thus shall pursue the question of origins by asking about those earlier thinkers whose work was decisive for Heidegger's development. Secondly, we shall consider origin in the sense of original or basic issue, and accordingly shall attempt to delimit this issue and to indicate how it serves as origin. Finally, we shall understand origin in its most radical sense as that which grants philosophical thought its content. At the level of this most radical sense of origin, it is perhaps possible for death to regain a signifying power for philosophy which, despite all differences, could match that which it had among the Greeks.

I. The Historical Origins of Heidegger's Thought

Taking origin, first, in the sense of historical origin, we ask: Who are those thinkers whose work served to set Heidegger's thought on its way? If, in posing this question, we let the concept of origins expand into that of mere influences, then the question proves right away to be unmanageable. With the exception of Hegel, no other major philosopher has so persistently exposed himself to dialogue with the tradition. And if we began to count up influences, even excluding all lesser ones, we would have to name Dilthey, Nietzsche, Kierkegaard, German Idealism, Kant, Leibniz, Descartes, Medieval Scholasticism, and Greek philosophy, that is, virtually the entire philosophical tradition--to say nothing of Heidegger's contemporaries nor of such poets as

Pindar, Sophocles, Hölderlin, and Trakl, all of whom were profound influences on Heidegger. Clearly such reckoning up of influences comes to nothing unless we first grasp the basic engagement of Heidegger's thought--that engagement on the basis of
which he is then led to engage in his extended dialogue with
nearly every segment of the tradition. Let us, then, pose our
question in a more precise and restricted way: What are the historical origins of the basic engagement of Heidegger's thought?
But the question is still inadequate. Engagement of philosophical
thought involves two moments: It is an engagement *with some
issue*, and it is an engagement with it *in some definite way*. In
other words, engagement involves both issue and method, and it
is of these that we need to consider the historical origins. Our
question is: What are the historical origins from which Heidegger
took over the issue and the method of his thought?

The method is that of phenomenology, and Heidegger took it
over from his teacher Edmund Husserl. It was for this reason
that Heidegger dedicated *Being and Time* to Husserl and therein
expressed publicly his gratitude for the "incisive personal guidance" that Husserl had given him. In various later autobiographi-
cal statements Heidegger speaks of the fascination which Husserl's
Logical Investigations had for him during his formative years and
of the importance which his personal contact with Husserl had for
his early development. In *Being and Time* phenomenology is explicitly identified as the method of the investigation; and in the
recently published Marburg lectures of 1927 Heidegger works
through almost the entire problematic of *Being and Time* under
the title "Fundamental Problems of Phenomenology."

What exactly did Heidegger take over from Husserl? What is
phenomenology? It is, in the first instance, the methodological
demand that one attend constantly and faithfully to the things
themselves. It is the demand that philosophical thought proceed
by attending to things as they themselves show themselves rather
than in terms of presupposed opinions, theories, or concepts.
Consider the case of perception. There have been theories which
claimed that, when we look at some object such as a desk, what
we really see are merely impressions or sensations in our own
mind. Within the framework of such a theory there can arise,
then, the almost insuperable problem of trying to explain how
such impressions confined within the mind can somehow be connected with real things outside the mind. This problem--a problem
arising from theory, not taken from the things themselves--can,
in turn, motivate the development of one or another version of

idealism intended to preserve some sense of objectivity. By contrast, phenomenology demands that such problems and the theories by which they are generated be put aside, set out of action. The phenomenological method involves, rather, attending to the perceptual object as it shows itself; it involves granting, against all constructivism, that what we perceive when we look at such an object as a desk is just the desk itself. It involves taking seriously, for example, the fact that perception of such objects is always one-sided, that we see only one profile of the desk at a time. It involves taking such a feature so seriously as to make it a basic determinant in the very concept of external perception.

Phenomenology is thus the appeal to the things themselves. And so, in *Being and Time* we find analyses such as that which Heidegger gives of tools. A tool, for instance, a hammer, normally shows itself within a certain context, namely, as belonging with other tools all suited to certain kinds of work to be done; only through a severe narrowing of perspective can we come to regard the hammer as a mere thing. Or, take the case of hearing; and consider: What sorts of things do we usually hear? We hear an automobile passing, a bird singing, a fire crackling-- whereas, as Heidegger says, "it requires a very artificial and complicated frame of mind to 'hear' a pure noise."[2] Yet, as a method, phenomenology extends beyond the sphere of things even in this enriched sense: Whatever the matter (*Sache*) to be investigated, the phenomenological method prescribes that it be investigated through an attending to it as it shows itself. Thus, *Being and Time*, dedicated primarily to the investigation of that being which we ourselves are and which Heidegger denotes by the word "Dasein," proceeds by attending to the way in which Dasein shows itself. What complicates the methodological structure of Heidegger's work is the fact that Dasein is also the investigator so that it becomes a matter of Dasein's showing itself to itself. Nevertheless, this complexity does not render the investigation any less phenomenological.

On the contrary, in that project to which his investigation of Dasein belongs, Heidegger seeks to be more phenomenological even than Husserl himself. He seeks to radicalize phenomenology by adhering even more radically than Husserl to the phenomenological demand to attend to the things themselves. As he expresses it in a later self-interpretation, he sought "to ask what remains unthought in the appeal 'to the things themselves'."[3] This dimension, tacitly presupposed in the phenomenological appeal to the things themselves, this dimension to which

Heidegger's radical phenomenology would penetrate, constitutes
the basic issue of Heidegger's thought.
What is this issue? What is fundamentally at issue in Heidegger's thought? One name for this issue--perhaps not the best--is
Being. This name betrays immediately the historical origin from
which Heidegger took the issue, namely, Greek philosophy, especially Plato and Aristotle. For it was in Greek philosophy that
Being was most explicitly and most profoundly put at issue, in
works such as Plato's *Sophist* and Aristotle's *Metaphysics*. Heidegger considers all subsequent reflections on Being, all later
ontology, as a decline from the level attained by the great Greek
philosophers: Gradually Being ceased to be held genuinely at
issue, and what Plato and Aristotle had accomplished, what they
had wrested from the phenomena, was uprooted from the questioning to which it belonged, became rigid and progressively
emptier. *Being and Time* is thus cast explicitly as an attempt to
raise again the question about Being. It is cast as a renewal, a
recapturing, of the questioning stance of Greek philosophy. This
is why it begins as it does: The very first sentence of *Being and
Time* is a statement not by Heidegger but rather by the Eleatic
Stranger of Plato's *Sophist*, a statement of his perplexity regarding Being. *Being and Time* literally begins in the middle of a
Platonic dialogue.
     Yet, on the other hand, *Being and Time* is no mere repeat-
ing of Greek philosophy. Heidegger does not seek to reinstate the
work of Plato and Aristotle, as though historicity could just be
set out of action in this exceptional case; nor does he propose
merely to revive the questioning in which their work was sustained. In his lectures of 1935, later published as *Introduction to
Metaphysics*, his intent is clear:

> To ask "How does it stand with Being?" means nothing less
> than *to recapture* [wieder-holen] the beginning of our histori-
> cal-spiritual Dasein, in order to transform it into a new
> beginning....But we do not recapture a beginning by reduc-
> ing it to something past and now known, which need merely
> be imitated; rather, the beginning must be begun again, more
> originally, with all the strangeness, darkness, insecurity that
> attend a true beginning.[4]

Heidegger would take up more originally the beginning offered by
Greek philosophy, take it up by taking it back to its sustaining
origin, make of that beginning a new beginning.

The historical origins of Heidegger's thought, in the restricted sense specified, are thus constituted by Husserlian phenomenology and Greek ontology. From the former Heidegger's method is taken; from the latter it receives its fundamental issue. However, method and issue are not simply unrelated. Rather, as we have already noted, Heidegger's penetration to what becomes the fundamental issue for his thought is, by his own testimony, an attempt to radicalize phenomenology, "to ask what remains unthought in the appeal 'to the things themselves.'" How is it that Being is what remains unthought in the appeal to the things themselves? How is it that a radical phenomenology must become ontology?

Let us consider again the characteristic approach prescribed by the dictum of phenomenology, "to the things themselves." What remains unthought here? What does the approach fail to take into account? The dictum prescribes that things are to be regarded as they show themselves. In thus attending to their showing of themselves--for example, in describing the one-sidedness with which external perceptual objects show themselves--what is lacking is attention to that which makes possible this showing. What remains unthought is the ground of the possibility of such showings as those to which phenomenology demands we attend.

But what kinds of grounds are relevant here? In proposing to attend to the grounds of the possibility of phenomena, is Heidegger proposing that we investigate the things in themselves in order to establish, independently of their self-showing in perceptual experience, how their physical constitution makes it possible for them to show themselves--how, for example, they are capable of reflecting light which then strikes our retinas so as finally to produce sensations in us? Clearly this is *not* what he is proposing. For such an extension would amount to a relapse into pre-phenomenological investigation rather than a radicalization of phenomenology. If the advance to the grounds which make it possible for things to show themselves is to remain phenomenological, if it is to constitute a radicalization of phenomenology, then those grounds must themselves be such as can be brought to show themselves.

Consider again the example of a tool. What is required in order for a hammer to show itself, not merely as a physical thing like the other physical things, but in its specific character as a hammer? It is required that it be linked up with a certain context of other tools, all oriented towards certain kinds of work to be done--especially if, as Heidegger insists, the hammer most genuinely shows itself as a hammer, not when I merely observe it disinterestedly, but rather at the moment when I take it up and use it for such work as it is suited to. In order for the hammer to show itself as a hammer (when I take it up and use it), there must be already constituted a context from out of which it shows itself--that is, a system of involvements or references by which various tools and related items belong together in their orientation, their assignment, to certain kinds of work to be done. Such a system of concrete references is an example of what Heidegger means by *world*.

So, what exactly is required in order that it be possible for the hammer to show itself as such? What is the ground of the possibility of the showing? It is required that there be something like a world from out of which the tool can show itself and, more significantly, that that world be *disclosed* in advance to the one to whom the tool would show itself. The hammer can show itself to me in that action in which I take it up and use it *only if* the system of involvements by which tools and tasks are held together is one with which I am already in touch, only if it is already disclosed to me. More generally, the ground of the possibility of things showing themselves is a prior disclosure of a world, of an open domain or space, for that showing. Radical phenomenology, as Heidegger pursues it, would penetrate to the level of such disclosure.

Still, however, it is not clear why radical phenomenology must become ontology. How is it that the investigation of such fundamental disclosure comes to coincide with a renewal of questioning about Being? This connection can be seen only if we consider with more precision just how Being is put at issue in *Being and Time*. What is asked about in the questioning of Being in *Being and Time*? It is the *meaning* of Being that is asked about. But what is asked about in asking about meaning? What is meaning? According to the analyses in *Being and Time* in which the concept of meaning is worked out, meaning is that from which (on the basis of which) something becomes understandable. To ask about the meaning of Being is thus to ask about how Being becomes understandable for us; it is to ask about Dasein's understanding of Being. Yet, understanding of Being is, in general, that which makes possible the apprehension of beings as such. Hence, to question about the meaning of Being, about Dasein's understanding of Being, is to ask about that understanding which makes it possible for Dasein to apprehend beings. It is to ask about that understanding which makes it possible for beings to show themselves to Dasein--that is, about that understanding which constitutes the ground of the possibility of things showing themselves. It is to ask about the opening up of the open space for such showing, about the disclosure of world, about *disclosedness*. To ask about the meaning of Being is to ask about Dasein's disclosedness.

We see, therefore, how ontological questioning and radical phenomenology converge in the basic problem of disclosedness. This matter of disclosedness is the fundamental issue. In it the issue and method which Heidegger takes over from his historical origins are brought together and radicalized. It is this issue, disclosedness, that can thus more properly be called the origin of Heidegger's thought.

II. The Original Issue of Heidegger's Thought

By thinking through the way in which Heidegger takes over his historical origins, we have come upon a second, more fundamental sense of origin, namely origin in the sense of original issue, the issue from which originates Heidegger's approach to other issues and his extended dialogue with the tradition. This issue is disclosedness.

In the various existential analyses in *Being and Time* it is readily evident that disclosedness is the original issue. For example, Heidegger's analysis of moods aims at exhibiting moods as belonging to us, i.e., to Dasein, in a way utterly different from the way in which redness belongs to a red object, different even from the way in which so-called inner states such as feelings have usually been taken to belong to man. He seeks to exhibit moods in their disclosive power, to exhibit them as belonging to Dasein's fundamental disclosedness. His analysis seeks to show that, among other functions, moods serve to attune us to the world, to open us to it in such a way that things encountered within that world can matter to us in some definite way or other-- in such fashion that, for instance, they can be encountered as threatening.

Heidegger's analysis of understanding is similarly oriented. Understanding is regarded not as some purely immanent capacity or activity within a subject but rather as a moment belonging to Dasein's disclosedness. Understanding is a way in which Dasein is disclosive. In understanding, Dasein projects upon certain possibilities, comports itself toward them, seizes upon them as possibilities; and from such possibilities Dasein is, in turn, disclosed to itself, given back, mirrored back, to itself. Dasein is given to understand itself through and from these possibilities. In addition, the possibilities on which it projects are disclosive in the direction of world, most evidently in the sense that they prescribe or light up certain contextual connections pertaining to the realization of the possibilities. When, for example, I project upon the possibility of constructing a cabinet, not only do I understand myself as a craftsman but also this possibility lights up and orients the context within the workshop.

Heidegger's analysis of death also remains within the compass of the issue of disclosedness, and indeed this is why it is so revolutionary. According to this analysis death is Dasein's ownmost and unsurpassable possibility; it is that possibility which is most Dasein's own in the sense that each must die his own death, and it is unsurpassable in the sense that Dasein cannot get beyond its actualization to still other possibilities; it is the possibility in which what is at issue is the loss of all possibilities. Heidegger's analysis focuses specifically on Dasein's comportment to this possibility, its projection on it, its Being-toward-death. Such projection is an instance of understanding, that is, it is a mode of disclosedness. In Being-toward-death Dasein is, in a unique way, disclosed to itself, given back to itself from this its ownmost possibility. Precisely because it is a mode of disclosedness, Dasein's Being-toward-its-end is utterly different, for instance, from that of a ripening fruit.

Thus, disclosedness is the original issue in Heidegger's analyses of Dasein. Through these analyses Heidegger seeks to display the basic ways in which Dasein is disclosive and to show how these various ways of being disclosive are interconnected. Indeed, not every basic moment displayed in the analyses of Dasein is simply a way of being disclosive. Yet even those structural moments that fall outside of disclosedness proper are still related to it in an essential way. More precisely, such moments are related to disclosedness in such a way that their basic character is determined by this relation.

Consider that moment which Heidegger calls "falling." This is the moment which he seeks to display through his well-known descriptions of the anonymous mass ("*das Man*")--his descriptions of how it ensnares the individual by its standard ways of regarding things and speaking about them; how it entices the individual into a conformity in which everything genuinely original gets leveled down and passed off as something already familiar to everyone; how, more precisely, it holds Dasein from the outset in a condition of self-dispersal and opaqueness to itself. What does this moment, this falling toward the rule of the anonymous mass, have to do with disclosedness? It has everything to do with it, because it is nothing less than a kind of counter-movement to disclosedness. It is a propensity toward covering up, toward *concealment*. This counter-movement toward concealment is essentially connected with Dasein's disclosedness. The connection is best attested by the issue of authenticity: Dasein's own genuine self-disclosure, the opening up of a space for its self-understanding, takes the form of a recovery of self from that dispersal in which the self and its possibilities are concealed beneath that public self that is no one and those possibilities that are indifferently open to everyone. Dasein must wrest itself from concealment.

Thus, Dasein's disclosedness is no *mere unopposed* opening of a realm in which things can show themselves. On the contrary, there belongs to that disclosedness an intrinsic opposition; there belongs to it a contention, a strife, between opening up and closing off, between disclosing and concealing.

Disclosedness, thus understood, is the original issue not only in the Dasein-analytic of *Being and Time* but also in Heidegger's later work. In order to grasp this continuity, we must consider a basic development which Heidegger's work undergoes after *Being and Time*. We should note, first, that already in the earlier work Heidegger brings the Dasein-analytic explicitly into relation with the problem of truth. He identifies the concept of disclosedness with that of truth in its most primordial sense; he presents disclosedness or original truth as constituting the ground of the possibility of truth in that ordinary sense related to propositions and the things referred to in propositions. Hence, the strife intrinsic to disclosedness may also be termed the strife of truth and untruth. For truth in this original sense, as that opening which provides the basis on which there can be true or false propositions regarding the things that show themselves in that opening--for truth in this sense Heidegger appropriates the Greek word *aletheia*.

In his later work Heidegger speaks of the original issue primarily in these terms, in terms of original truth or *aletheia* instead of disclosedness. And, though the issue remains the same, there is, nevertheless, behind this shift in terminology a fundamental development. What is the development? It may be regarded as a progressive separation of two phenomena which in *Being and Time* tended to coalesce. Specifically, Heidegger comes in the later work to dissociate truth from Dasein's self-understanding--that is, he dissociates the contentious opening up of a realm in which things can show themselves (i.e., truth) from the movement of self-recovery by which Dasein is given to itself. The happening of truth is set at a distance from the reflexivity of human self-understanding. But original truth remains for Heidegger the original issue; and so, with its dissociation from self-understanding, Heidegger's thought is set more decisively at a distance not only from German Idealism and the tradition which led to it but also from that idealistic path which Husserl himself followed in his later work.

Granted this development, the original issue of Heidegger's thought remains in the later work what it was from the beginning, namely, the opening up of a domain in which things can show themselves--that is, the issue of original truth. Consider, for example, Heidegger's essay on the work of art. In this essay Heidegger opposes the modern tendency, stemming from Kant, to refer art to human capacities such as feeling that could be taken as having no connection with truth; contrary to such an approach, Heidegger seeks to show that original truth is precisely what is at issue in art. According to his analysis a work of art makes manifest the strife of truth. It composes and thus gathers into view truth in its tension with untruth. A work of art presents the strife between world, i.e., the open realm in which things can show themselves, and earth, i.e., the dimension of closure and concealment.

Heidegger's analysis of technology in his later works is similarly oriented. This analysis, which is something quite different from a sociological, political, or ethical reflection on technology, is directed, strictly speaking, not at technology as such but rather at what Heidegger calls the essence of technology. What is the essence of technology? It is simply a mode of original truth, the opening of a realm in which things come to show themselves in a certain way. It is, specifically, that opening in the wake of which nature comes to appear as a store of energy subject to human domination. It is that opening in which natural things show themselves as to be provoked to supply energy that can be accumulated, transformed, distributed, and in which human things show themselves as subject to planning and regulation. What is at issue in Heidegger's analysis of technology is that same original issue to which his thought is already addressed from the beginning. It is that issue in which converge his efforts to radicalize Husserlian phenomenology and to renew Greek ontology, the issue of disclosedness, of original truth.

III. The Radical Origin of Philosophical Thought

There is still a third sense of origin which we need, finally, to bring into play. This third sense is not such as to revoke what has been said regarding truth as the origin of Heidegger's thought. It is not a matter of discovering some origin other than truth but rather of deepening, indeed radicalizing, the concept of origin. It is a matter of grasping truth as radical origin.

In order to see how this final sense of origin emerges, we need to grasp more thoroughly the methodological character of the analyses of Dasein in *Being and Time*. Contrary to what might seem prescribed by the phenomenological appeal to the things themselves, Heidegger's analyses are not simply straight-forward descriptions of Dasein as it shows itself. Why not? Because ordinarily Dasein does not simply show itself. Rather, there belongs to Dasein a tendency toward self-concealment of the sort that Heidegger discusses, for example, in his analysis of falling. What does this entail as regards the method required of a philosophical investigation of Dasein? It entails that Dasein, rather than being merely, straightforwardly described, must be wrested from its self-concealment.

But how, then, we must ask, is the investigation to be freed of the charge of doing violence to the phenomena? How can such investigation claim to be phenomenological? How can it justify the claim of proceeding solely in accord with the manner in which the things themselves show themselves? There is only one way. The violence that is done must be a violence which Dasein does to itself rather than a violence perpetrated by the philosophical investigation. The wresting of ordinary Dasein from its concealment must be the work, not of a philosophical analysis which would inevitably distort it and impose on it something foreign, but of a latent disclosive power within Dasein itself. Heidegger is explicit about the matter: The philosophical analysis must, as it were, "listen in" on Dasein's self-disclosure; it must let Dasein disclose itself, as, for example, in anxiety. Attaching itself to such disclosure, the philosophical analysis must do no more than merely raise to a conceptual level the phenomenal content that is thereby disclosed.

This peculiar methodological structure is what determines the final sense of origin. How? By virtue of the fact that it simply traces out the connection of thought to its sustaining origin. More specifically, this structure prescribes that Dasein's self-disclosure is precisely what gives philosophical thought its content, what grants it, yields it up to thought. Dasein's self-disclosure, that self-disclosure on which philosophical thought "listens in," is thus the origin of that thought--not just in the sense of being the central theme for that thought but rather in the sense of first granting to such thought that content which it is to think. Yet, Dasein's self-disclosure is simply a mode of Dasein's disclosedness as such--that is, a mode of original truth. Truth is what grants to thought that content which it is to think. The origin of thought is original truth.

The genuine radicalizing of the concept of origins comes, however, only in the wake of the development that takes place in Heidegger's later work. Within the framework of *Being and Time* there is no difficulty involved in understanding how philosophical thought can attach itself to its origin, to original truth; for such truth, though perhaps merely latent, is not essentially removed from Dasein. Philosophical thought can attach itself to its origin, because that origin belongs latently to each of us, including whoever would philosophize. We are always already attached to original truth. The problem arises when, through the experience of the history of metaphysics, Heidegger comes in his later work, to dissociate truth from self-understanding. For this amounts to placing original truth at a distance from Dasein--that is, at a distance from that thought whose origin that truth would be. Thus, Heidegger's later work has to contend with a separation between original truth and that thought to which it would grant what is to be thought. As a result, the granting becomes a problem. Truth, the origin of thought, essentially withdraws from thought, holds itself aloof. Truth is the *self-withdrawing origin of thought*. And thought, resolutely open to the radical concealment of its origin, lets itself be drawn along in the withdrawal. Here we arrive at the most radical sense of origin.

Heidegger's efforts to radicalize Husserlian phenomenology and to renew Greek ontology converge on truth, first, as the original issue or basic problem and then, finally, as the origin which grants philosophical thought as such. What is most decisive in this most radical concept of original truth is that truth so conceived withdraws from that very thought which it grants and engages. It withholds itself from thought.

What is remarkable is that the same may be said of death. It too withholds itself from thought, withdraws from every attempt to make it something familiar. In distinction from all other possibilities, death alone offers us nothing to actualize in imagination.

It offers us no basis for picturing to ourselves the actuality that would correspond to it. It is sheer possibility, detached from everything actual, detached from us, self-withdrawing--yet constantly, secretly engaging.

Death withdraws as does original truth--withdraws while yet engaging us. And so, death has the power to signify original truth. Yet, the task of philosophy, the task to which Heidegger finally came, is to develop thought's engagement in such truth. And so, death, signifying original truth, signifies the end to which philosophy is directed. At this level death can become a positive symbol for philosophy.

Perhaps it is a more fitting memorial to Heidegger if, instead of merely dwelling on his death, we seek to restore to death its power to signify the end and thus the task of philosophy.

NOTES

1. The address is published in *Max Scheler im Gegenwarts-geschehen der Philosophie*, ed. P. Good (Bern: Francke Verlag, 1975).

2. *Sein und Zeit* (Tübingen: Max Niemeyer Verlag, 1963), p. 164(207).

3. *Zur Sache des Denkens* (Tübingen: Max Niemeyer Verlag, 1969), p. 71(64).

4. *Einführung in die Metaphysik* (Tübingen: Max Niemeyer Verlag, 1958), pp. 29-30; *An Introduction to Metaphysics*, trans. Ralph Manheim (Garden City: Doubleday, 1959), p. 32.

6. JOHN D. CAPUTO

HUSSERL, HEIDEGGER AND THE QUESTION OF A "HERMENEUTIC" PHENOMENOLOGY

In § 7c of *Being and Time* Heidegger writes, "The phenomenology of Dasein is a *hermeneutic* in the primordial signification of this word, where it designates this business of interpreting" (SZ, §7c, 37/62).[1] Yet just a few lines earlier in the same section Heidegger had written that the notion of a "descriptive phenomenology" is "at bottom tautological" (SZ, § 7c, 35/59). But if phenomenology is inherently descriptive, how can it be at the same time interpretive? Phenomenology deals with what is given, while hermeneutics means taking something *as*, construing it, giving it a "rendering." Moreover, phenomenology has to do with the things themselves, *die Sachen selbst*, not with philosophers and their philosophies,[2] while hermeneutics has to do with texts, with writings. One can at least imagine how Derrida's famous motto, "*il n'y a pas de hors texte*," might have arisen emerged from hermeneutics, but one can never imagine it arising in a Husserlian framework.

How then can phenomenology take a hermeneutic turn? Is hermeneutic phenomenology an aberration, a commingling of a foreign substance in the organism of phenomenology? Or does it represent an extension of something essential in the phenomenological method? Put in historical terms, how is it possible to move from the "pure" phenomenology of Husserl's *Ideas I* to the hermeneutic phenomenology of *Being and Time*? That is the question I want to pose in this paper, and the argument I will make is that there is already a distinctively hermeneutic element in pure phenomenological investigation (even as I would want to insist that there must be a distinctively phenomenological element in all hermeneutics, lest it fall prey to the textualism of Derrida and the deconstructionist critics in French and American literary criticism.) I want to show that Heidegger's phenomenology results

104

from an appropriation of a hermeneutic element which is already at work in Husserl himself. If *Being and Time* practices a hermeneutic phenomenology, this is because Heidegger has acted upon certain suggestions of Husserl, exploited certain resources in Husserl's own method, moved phenomenology in a direction which Husserl himself made possible. If the phenomenology of Heidegger is explicitly hermeneutic, Husserl's phenomenology is already in an important sense a "proto-hermeneutics."

It would be well at this point to say what hermeneutics means, for like so many words whose day has come, which enjoy such currency and wide circulation, "hermeneutics" is in danger of meaning almost anything. For the purposes of the present study, I take the word to refer to the theory of the fore-structures, and to the notion that the understanding of an entity is possible only insofar as it is made possible by a certain prior projection of the range or field of meaning to which the entity belongs. Hermeneutics thus turns on the famous hermeneutic circle: If we do not already stand in a certain prior understanding of the thing to be understood, then understanding is disoriented, cut adrift and hence rendered impossible. Hence my tactic here is to take the Heideggerian sense of hermeneutics as the orthodox notion and then to see to what extent it is both prepared for by, and hence in the debt of, Husserl's phenomenology.

Now there is no little paradox in such an inquiry. For Heidegger's hermeneutic phenomenology is commonly set in opposition to Husserl's pure phenomenology. Husserl, it is usually pointed out, wanted phenomenology to be a presuppositionless science, while Heidegger said that it is not at all a question of getting rid of our presuppositions but rather of finding the right ones, the ones which illuminate the things themselves. It is not a question of getting out of the circle, but of getting into it in the right way. But I want to argue that this is far too facile a way to differentiate Heidegger and Husserl. The fact of the matter is that the fore-structure of the understanding, which belongs to the essence of Heidegger's hermeneutical method, plays a central role in Husserl as well. I want to show that Husserl depended upon such a notion at every critical juncture in his account of perception, that it is a centerpiece of his phenomenology to which he devoted numerous and typically painstaking analyses. In the conclusion of the paper I will address the question of the real difference between Heidegger and Husserl and of the validity of the common wisdom about the place of presuppositions in their respective phenomenologies.

I will carry out this undertaking as follows. (1) I will take my point of departure from Husserl's discussion of "explication" (*Auslegung*) in the *Cartesian Meditations* in order to show that it bears an important relationship to Heidegger's use of the same word in *Being and Time*; (2) I will then turn to a question of the anticipatory "predelineation" (*Vorzeichnung*) which Husserl treats in *Ideas I*. (3) Next I take up Heidegger's hermeneutic phenomenology in *Being and Time* and its relationship to these elements in Husserl's work. (4) And then in conclusion I return to the question of presuppositionlessness and address the question of the true divide between the Husserlian and Heideggerian programs.

THE MEANING OF "AUSLEGUNG" *IN THE* "CARTESIAN MEDITATIONS"

It often goes unnoticed that the word which Dorian Cairns translated as "explication" in his English version of *Cartesian Meditations*, and which Macquarrie and Robinson render as "interpretation" in their translation of *Being and Time*, is in both cases "*Auslegung*." While each of these translations seems to me contextually fair, this should not obscure the convergence of the Husserlian and Heideggerian vocabulary at this point.

Husserl introduces the word into the *Cartesian Meditations* in § 20 in a discussion of "intentional analysis," which is to be distinguished from analysis in the usual sense as follows:

It becomes evident that, as intentional, the analysis of consciousness is totally different from analysis in the usual and natural sense. Conscious life, as we said once before, is not just a whole made up of "data" of consciousness and therefore "analyzable" (in an extremely broad sense, divisible) merely into its self-sufficient and non–self-sufficient *elements*... (CM, § 20, 83/46)[3]

Conscious life is not made up of atomic data which can be separated out and then recombined. Rather it functions in accord with a law of its own, which Husserl identifies as the law of the implicit and the explicit. The passage continues:

...but everywhere its peculiar attainment (as "intentional") is an uncovering of the *potentialities* "*implicit*" in actualities of consciousness--an uncovering that brings about, on the noematic side, an "explication" or "unfolding" (*Auslegung*), a "becoming distinct" and perhaps a "clearing" (*Klärung*) of what is consciously meant (the objective sense) and, correlatively, an explication of the potential intentional processes themselves.

Intentional experience is not a composite of smaller, and isolatable "actualities," but rather an interplay between a focal actuality and a ring of potentiality, a ring of implicit structures surrounding what is thematic and explicit. The intentional object is not a merely present actuality--Heidegger would say, not merely something *vorhanden*--rather it is "more" than it actually appears to be. That means that the task of analyzing, which must be tailored to the unique mode of being of the intentional object, must consist not only in attending to the focal object but just as importantly in unfolding these implicit structures, in laying out or setting forth the muted components of intentional life which nonetheless belong integrally to its make-up. And that is the precise force of the word *Auslegung*: to lay out, to set forth, somewhat in the manner of the Latin *ex-ponere*. Cairn's "explicate" captures the Latin *explicare, explicatus*: to un-fold, to lay out, the plies or folds which are folded together in the *explicandum*. The work of phenomenological reflection then is to make the implicit explicit, to show the dependence of the thematic object upon the pre-thematic and implicit components. In this sense, phenomenology is an *ars explicandi*.

This notion that the intentional object is always "more" than it first presents itself to be casts an interesting light on the principle of principles (*Ideas I*, § 24). For to accept a thing just as it presents itself to be and not otherwise has nothing to do with a shortsighted and narrow empiricism which constricts experience into abbreviated atoms of experience. The principle of principles must always be qualified by Husserl's observation in the *Analyses of Passive Synthesis* about the pretentions of perception:

And since perception still pretends to give the object in person (*leibhaft*), it continually pretends in fact more than it can do in accordance with its own essence.[4]

The intentional object always contains a *plus ultra*, something more, which it is the task of phenomenological reflection to unfold and elaborate:

Intentional analysis is guided by the fundamental cognition
that, as a consciousness, every cogito is indeed (in the
broadest sense) a meaning of its meant (*Meinung seines
Gemeinten*), but that, as a consciousness at any moment, this
something meant (*dieses Gemeinte*) is more--something meant
with something more--than what is meant at that moment
"explicitly." In our example, each phase of perception was a
mere side of "the" object, as what was perceptually meant.
This *intending-beyond-itself*, which is implicit in any consciousness, must be considered an essential moment of it.
(CM, § 20, 84/46)

Whence the Derridean "deconstruction" of Husserl's phenomenology
misconstrues what the phenomenological method demands of itself.
It does not identify the perceptual object with pure presence, nor
does it expect that perception can deliver it. Rather it insists all
along that the intentional object is girded about with a ring of
potential and implicit structures which it is up to phenomenological analysis to unpack.

 Husserl of course has in mind his well known doctrine of the
inner and outer "horizons" which show the contextualized make-up
of perception and of intentional experience generally.[5] Husserl
never regarded perception as a simple matter of "looking on" as if
intuition (*An-schauung*) were merely a blank stare. On the contrary, from the *Logical Investigations* on, he held that *Anschauung* is *Auffassung*:[6] that intuiting is construing, apprehending,
even interpreting. To be able to perceive something demands of
the perceiver that he know how to *take* the perceptual object,
know how to contextualize and situate it, so that it can appear *as*
the kind of thing which it is. If we are not alert to the fringe of
potential and implicit components in the intentional object, our
perception of it will simply break down. That is the point of
Husserl's often repeated example of the *Abschattungen*, "adum-
brations" (which literally means "fore-shadowing"). The perceptual object is not directly perceived, but rather ap-perceived
along with its co-given, mediately presented fringe of inner and
outer horizons. Were we not alert to the potential factors in what
is directly presented, the perceptual object would be reduced to a
façade and the encounter with its other sides would be a dis-
continuous and indeed surprising addition to a fragmentary
experience. Perceptual experience thus depends upon having a
grasp of these potential factors, for these alone allow us to *construe* the object rightly, in a manner which befits its appearance.

Intentional analysis consists in the hard work of setting forth, of setting out (aus-legen), these potential factors, factors which, as Heidegger often points out, are all the more difficult to seize precisely because they are implicit.

Thus it begins to become clear that if Husserl subscribed to the idea of presuppositionless science, this has nothing to do with his idea of perception, and nothing to do with the myth of uninterpreted bits of intentional data which are independent of the horizons and potentialities which structure and condition their appearance.

Now the make-up of these horizons merits further consideration. For the horizonal potentialities are not wholly indeterminate, or merely "absent" to use the language of the day, but rather, as Husserl puts it "predelineated" (vorgezeichnet): "The horizons are predelineated potentialities" (CM, § 19, 82/45). Husserl's choice of words here is telling. Vor-zeichnen means to sketch something out beforehand, to lightly trace it out in advance, as in a drawing, say, where the focal object is done in detailed and heavy strokes while the background is simply traced in lightly. The background-horizon is neither there nor not there, neither present nor absent, but rather something adumbrated or foreshadowed. Hence in Derridean terms it is a trace but a trace which, pace Derrida, prepares us for the presentation of what is on the margin. Whence the words Vorzeichnung and Abschattung are importantly akin and akin precisely inasmuch as both indicate a certain fore-structure, referring as they do to a fore-sketching, fore-shadowing or preliminary tracing out. Husserl also uses the Vor-deutung in another text, which Cairns recommends be translated as "preliminary indication," but which of course can also have the sense of preliminary interpretation. In Erste Philosophie Husserl refers to the outer horizon as a sphere of Vordeutung, which refers us to nearby or coexistent entities, all of which belong to a realm of possible experience.[7]

These are pregnant and suggestive words for the present study inasmuch as they open up the question of the fore-structure of intentional consciousness in Husserlian phenomenology. The cogitatum is possible only inasmuch as it has been anticipated in advance, only insofar as an adequate preparation has been made for it. There is an intrinsically anticipatory structure to intentional life. Consciousness does not pass from one intentional object to another as through a series of discrete objects, but rather it moves ahead anticipatorily into objects whose appearance has been announced in advance. The forward progress of consciousness is accordingly not a matter of entering a wholly new world, but rather of filling in (*Erfüllung*) a predelineated sketch. [8]

It is likewise clear that the ultimate fore-structure for Husserl is the temporal structure of pro-tention, the thrust of consciousness by which it tends forth into what lies immediately ahead in the concatenation of experience. But it is important to see that what is protended belongs to the immediate and present structure of experience, that consciousness, in other words, is already stretched out anticipatorily into the future. Otherwise Husserl's entire analysis would be frustrated and the role of the horizonally predelineated in the constitution of present experience undermined. Husserl's point is that intentional life does not consist of atomic, merely present actualities, but is precisely made possible by the fringe of horizonal potentialities, both inner and outer, both retained and protended, which belong to its integral structure.

THE DOCTRINE OF "PREDELINEATION" IN "IDEAS I"

I have thus far singled out for emphasis two notions in Husserl's later work which, I wish to argue, are essential ingredients in the later development of a hermeneutic phenomenology, viz., the notion of intentional explication on the one hand and of projective predelineation on the other. Now it is clear that these are correlative notions. Projective predelineation is the actual, prereflective work of consciousness, *in actu exercitu*. It is the rule which transcendental consciousness follows in its anonymous life of constituting objects. The notion of intentional explication, on the other hand, refers to the act of reflection which sets about unfolding and rendering explicit the muted and potential structures which are effective whenever any object becomes focal and thematic, raising these structures up to the level of reflective clarity, *in actu significato*. Projection and explication thus are related as the prereflective and the reflective, the anonymous and the explicit, the constitutive and the reconstructive. Explication wants to retrace the steps of projective predelineation, even as transcendental reduction retraces the steps of transcendental constitution.

I want now to pursue further the theory of projective "predelineation" by following Husserl's celebrated discussion of the destructibility of the world in §§ 47-49 of *Ideas I*, for this text throws important light upon Husserl's views. One should however dispel at once the unfortunate and misleading Cartesian overtones of this passage. For it is not Husserl's intention to show that consciousness enjoys a privileged metaphysical autonomy such that, were the physical world annihiliated, it would continue to exist.[9] On the contrary, Husserl's intentions are entirely epistemic. He is bent upon making a point about the nature of constitution, a point which, it seems to me, develops further his insight that the projective predelineation or anticipation of the intentional object is an essential condition of the possibility of its appearance. And having established that point, it will not be difficult to show the dependence of Heidegger's analysis of *Verstehen* and *Auslegung* in *Being and Time* upon Husserl's theories.

The pivotal term in this discussion is "motivation." Husserl wants to show that an object is never something in itself, without relationship to consciousness, but always the correlate of the conscious acts in which it is built up or constituted. But an object is constituted only inasmuch as consciousness is motivated to constitute it, that is, moved by regular concatenations of experiences to pick out particular formations which it endows with objectivity. The walking stick takes shape in our experience only because of the dependability of its properties, its long, thin, hard and supportive substance. Were it now hard and now soft, now long and now short, were it suddenly to become too hot to hold, or suddenly to drain into a pool of boiling liquid, then "it" would never be constituted; there would remain only a chaos of sensations over which consciousness would be left to exert an uneasy rule. An object is constituted only inasmuch as consciousness is motivated to constitute it by the unity and the stability of its sense.

This discussion leads Husserl to make an interesting distinction in § 48 between two species of possibility. The world of actualities is the world for which consciousness has been provided actual concatenations of experiences or actual motivation. It is the world which has actually taken shape in experience. Possibility, on the other hand, refers to the sphere of things whose possibility has been opened up by the actual concatenations of experiences. And that means that the possible in the primary sense means the experienceable (*erfahrbar*), which is to be distinguished from merely empty, logical possibility:[10]

Experienceableness never means a mere logical possibility, but rather a possibility *motivated* in the concatenations of experience. This concatenation itself is, through and through, one of "motivation," always taking into itself new motivations and recasting those already formed. (*Ideas I*, § 47, 89/106-107)

And motivations are more or less determinate, are possessed of greater or lesser definition, precisely in proportion to the extent to which consciousness has been moved to form them. Some possibilities are extremely close, while others may be quite remote, like the possibility of a rock at the bottom of the sea which no eyes have yet seen or constituted. Such a thing, Husserl would point out, has been only roughly pre-figured and predelineated by the actual experience which we have had up to now. But it is clear that Husserl is not here flirting with some kind of metaphysical idealism, which would deny the natural being of the rock on the grounds that it has not been constituted. On the contrary, it is only because we are presently able to project the horizon within which such an entity could make an appearance that it is indeed a real and not merely an empty possibility.

...anything whatever which exists in reality but is not yet actually experienced can become given and this means that the thing in question belongs to the undetermined but *determinable* horizon of my experiential actuality at the particular time. (*Ideas I*, § 47, 89/107)

The possible object in the primary sense of possible is the object whose appearance has been prepared for or "construed in advance" (im *voraus "aufgefasst*," CM, § 19, 83/45) of its appearance. It is not an empty, merely logical possibility but a concrete, motivated possibility, one which has been predelineated with respect to what Husserl calls its "essential type" (*Wesenstypus, Ideas I*, § 47, 90/107). This predelineated type is the phenomenological a priori which renders future experience possible, and indeed which renders present experience possible inasmuch as it too was prepared for on the basis of a prior experience. This is no Kantian a priori, however, for it is an a priori which has been drawn from or motivated by experience, and which is subject to ongoing temporal revisions, as former present horizons are filled or frustrated and new ones opened up.

Such concrete, predelineated possibilities are then entirely different in kind from the merely empty, logical possibility of, say, a world which would be entirely different than ours, which would be possible only in the sense that it involves no formal contradiction. For it is impossible to say anything more about such a world, for which we have no projective interpretation, no motivated predelineation. It is interesting to see, then, that Husserl in fact holds to something like a "fusion of horizons" theory. He insists that if the notion of something "other"-- another form of life, say, or another world--is to make any sense, it must be possible in principle to locate such a thing somewhere on the outer horizonal limits of actual experience. He writes:

> ...what is cognizable by one Ego must, of *essential necessity*, be cognizable by *any* Ego. Even though it is not *in fact* the case that each stands, or can stand, in a relationship of "empathy," of mutual understanding with every other, as, e.g., not having such relationship to mental lives living on the planets of the remotest stars, nevertheless there exist, eidetically regarded, *essential possibilities of effecting a mutual understanding* and therefore possibilities also that the worlds of experience separated in fact become joined by con-catenations of actual experience to make up the one intersub-jective world... (*Ideas I*, § 48, 90/108)

There must be a certain point of assimilability, that is, of a pre-delineation which, in however an inadequate way, prepares us to understand a novel experience, after which, of course we would then be in a position to revise our subsequent expectations. Whence a dialogue with an extra-terrestial consciousness must be prepared for by actual experience, that is, by assimilating the objects of its experience to the horizons which are familiar to us, while at the same time revising these horizons on the basis of the novelties of the other experience. Thus gradually, by a back and forth movement between assimilation and revision, carried out on both sides of the dialogue, a chain of experiences would be con-stituted which would link one experience with the other within the unity of a single world.

And that makes non-sense, or what Husserl here calls quite precisely "material counter-sense" (*sachlicher Widersinn*), out of the hypothesis of a world of experience which is entirely hetero-geneous to ours. Such a world is not a formal absurdity (*Unsinn*), the way a square circle is a contradictory notion, but it is a material counter-sense inasmuch as it makes no real, material, substantive sense. Our powers of projective thought, of horizonal predelineation, break down before it and we find

ourselves confronted with a merely "transcendental object =X," as
Kant put it, an object to which we can give no possible shape,
which we cannot prefigure, trace or sketch out beforehand in
even the remotest way, by means of even the lightest traces:

> When that is taken into account, the formal-logical possibility
> of realities outside the world, the *one* spatio-temporal world,
> which is *fixed* by our actual experience, materially proves to
> be a counter-sense. If there are any worlds, any real physi-
> cal things whatever, then the experienced motivations consti-
> tuting them must be *able* to extend into my experience and
> into that of each Ego in the general manner characterized
> above. (*Ideas I*, § 48, 90-91/108-109)

Husserl also links his theory of preobjective predelineation
and of explicative analysis to a theory of actual and potential
consciousness. Intentional experience is not a whole composed of
actualities, but a composite of the actual and the potential, that
is, the focal and the horizonal. While consciousness is actually
turned towards (*zugewendet*) an object, there remains a whole
field of potential or inactual (*inaktuell*) consciousness, that is, of
what is potentially or implicitly intended. Actual intentions are
surrounded by a "halo" of potential ones. The actual, waking ego
is ensconced by a host of potential intentionalities (*Ideas I*, 35).
For there are always lines reaching out from the thematic object
to the field to which it belongs, even as there are, on the side of
the subject, acts to which the ego can switch in order to activate
these possibilities. The ego can always alter the ray of mental
attentiveness, the beam of intentional regard (*Blickstrahl, Ideas
I*, § 92, 189/222), away from the object and towards the horizon,
even as it can alter its own modality from perception to memory
or imagination. It can bury itself in a matter of urgent concern
or disengage itself reflectively from it in order to survey its
properties (§ 50).

In § 115 Husserl expands the notion of an act to include a
distinction between acts which are actually carried out (*vollzog-
ener Akt*) and acts which are merely "stirring" (*Aktregung*). The
cogito in the proper sense is explicit intentionality, that in which
the ego actually lives, but this is to be distinguished from acts
which are alive before we know it:

> E.g. a belief, an actual belief, is "aroused" (*regt sich*); we
> already believe "before we know it." Similarly, under circum-

stances, positing of likings or dislikings, desirings, even resolves, are already alive before we "live" in them, before we effect the *cogito* proper, before the Ego is activated judgingly, likingly, desiringly, willingly. (*Ideas I*, § 115, 236/273)

The cogito in the proper sense is the explicit intentionality of wakeful consciousness, but the notion of conscious act must be broad enough to include non-explicit, pre-thematic acts. And that of course is a significant expansion of the term, for it confirms what we have all along been arguing, that the principle of principles has nothing to do with restricting ourselves to the momentary, the actual, the present. Rather it enjoins us to understand both object and act in the broadest sense, and hence to take into account the potential, implicit and tacit structures which are at work in conscious life. We already live in certain beliefs, we are already cognizant of certain objects, long before they are picked out by the searching ray of attentive regard. And that means that the work of phenomenology for Husserl is essentially a task of *Auslegung*, of laying-out and unfolding, the implicit dimensions which make explicit intentionality possible.

HEIDEGGER'S HERMENEUTIC PHENOMENOLOGY

I want now to show that Heidegger's hermeneutic phenomenology turns upon Husserl's discovery of the sphere of phenomenological fore-structures. Husserl's notion of a predelineated horizon is of decisive importance to Heidegger's notion of projective understanding (*entwerfendes Verstehen*); the Husserlian *Vorzeichnung* makes the Heideggerian *Vorstruktur* possible. Hermeneutical "interpretation" (*Auslegung*) has a point of departure in intentional explication (*Auslegung*). Hence we want to show that Husserl's phenomenology is at bottom already a proto-hermeneutics, already a philosophy of the presuppositions which infiltrate and condition thematic acts. And that is at least a surprising thing to say of this philosopher of presuppositionlessness, and underlines the question which we raised at the beginning of this study--about the true line of separation between Heidegger and Husserl--to which we will return in the conclusion of this essay.

Heidegger's employment of a "hermeneutic" phenomenology in *Being and Time* went hand in hand with the "existential" make-up of the being which he called Dasein. He makes this plain in the highly significant methodological reflections contained in §63. Hermeneutic phenomenology, he writes there, must have the character of a projection of the primordial Being of Dasein which cuts through the concealments under which Dasein's Being is buried by the "public interpretation" of Dasein. And this projection, he says, is necessarily of a violent character:

> Dasein's *kind of Being* thus *demands* that any ontological Interpretation which sets itself the goal of exhibiting the phenomena in their primordiality, *should capture the Being of this entity, in spite of this entity's own tendency to cover things up*. Existential analysis, therefore, constantly has the character of doing violence (*Gewaltsamkeit*), whether to the claims of the everyday interpretation, or to its complacency and its tranquillized obviousness. While indeed this characteristic is specially distinctive of the ontology of Dasein, it belongs to any interpretation, because the understanding which develops in Interpretation has the structure of a projection. (SZ, § 63, 311–12/359)

For Heidegger hermeneutic projection is the counter-movement of existential fallenness. Dasein tends, by the natural momentum of fallenness, to drift outside of a proper self-understanding, and to interpret itself in the light of the public understanding. It tends to construe its most profound personal responsibilities in terms of the tranquillized obviousness of what "they" say. The task of hermeneutics, as Heidegger sees it, is to counter this existential drift, to reverse this movement, and hence to break the hold of fallenness:

> The laying-bare of Dasein's primordial being must rather be *wrested* from Dasein by following the *opposite course* from that taken by the falling ontico-ontological tendency of interpretation. (SZ, § 63, 311/359)

Thus we see how Heidegger has enlisted a fundamentally Husserlian notion, that to understand something is to know how to project it in the right terms, in the service of an existential and Kierkegaardian program: to reverse the movement of fallenness and to recover Dasein in its authentic Being. Let us investigate now just how the notion of projection and fore-structure function in *Being and Time*. Then we will be in a position to contrast the Heideggerian and Husserlian notions.

"Projection" (*Entwurf*) is the fundamental structure of understanding (*Verstehen*) in *Being and Time*: to understand something is to project it aright, that is, in terms of the horizon which is appropriate to it. Whence the Husserlian notion of horizon is equivalent for Heidegger to the Being, or mode of Being (*Seinsart*), of the entity. By means of its projective understanding Dasein is able to "cast" a being in a certain light, to situate it within a suitable framework. The English "cast" seems to be the right word here, for it operates within the rule of the same metaphor: to throw or thrust a thing into a certain horizonal framework (*werfen* = cast, hurl). The first task of *Being and Time* then is to make the right projective move with respect to Dasein, to cast Dasein in the appropriate terms, to project this being upon the Being which lets it be what it is. But this raises the further question as to how we are to settle upon the appropriate projection. Heidegger himself poses this problem with all desirable clarity:

> Where are ontological projects to get the evidence that their findings are phenomenally appropriate? Ontological Interpretation projects the entity presented to it upon the Being which is that entity's own, so as to conceptualize it with regard to its structure. Where are the sign posts to direct the projection, so that Being will be reached at all? (SZ, § 63, 312/359)

By what is ontological projection "*guided* and *regulated*" (ibid.)? How are we to know when we have seized an entity in its primordiality?

The answer to this question is disconcerting in its simplicity: we are *all along* possessed of the appropriate understanding of Dasein; this is something which we already understand, even if we have never conceived it and put it into words. Hence it is at this crucial juncture that Heidegger invokes the dynamics of the hermeneutic circle, which means that he will in a certain sense "presuppose" that which is to be demonstrated by the hermeneutic analysis. And it is at this point also, I want to insist, that his hermeneutical circle takes on a profoundly Husserlian sense. Heidegger's argument is that we can achieve primordiality only if we are already possessed of a primordial understanding. Hence the task of his hermeneutic phenomenology is not to propose something new but to explicate, bring to the surface, or lay out (*auslegen*) an understanding with which we are already equipped. We always and already move about within an understanding of the

Being of Dasein, and indeed of Being at large (SZ, § 2, 5/25), and the task of hermeneutic phenomenology is to raise this pre-understanding up to the level of an ontological concept. Hence *Being and Time* opens with a crucial decision: to project Dasein upon the horizon of "existence" (SZ, § 4, 12/32-33), which is for Heidegger nothing more than making an explicit concept of an understanding in which we all already subsist. The circularity of the hermeneutic circle is innocent of any formal fallacy, for it refers to the movement from an implicit understanding to an explicit concept. All the dilemmas surrounding the hermeneutic circle are solved in principle by recognizing the role of this simple, and quite Husserlian strategy, viz. of the logic of implicit and explicit. It is interesting to me that this same point has been noticed by Jacques Derrida, although for reasons of his own (misguided ones in my opinion), he takes this to be a criticism of both Heidegger and Husserl:

> ...the process of disengaging or of elaborating the question of Being, as a question of the *meaning* of Being, is defined as a *making explicit* or as an interpretation that makes explicit. The reading of the text Dasein is a hermeneutics of unveiling or of development (see sec. 7). If one looks closely, it is the phenomenological opposition "implicit/explicit" that permits Heidegger to reject the objection of the vicious circle, the circle that consists of first determining a being in its Being, and then of posing the question of Being on the basis of this ontological predetermination (p. 27). This style of a reading which makes explicit, practices a continual bringing to light, something which resembles, at least, a coming into consciousness, without break, displacement or change of terrain.[11]

The hermeneutic strategy of *Being and Time* is to tap into the operative, pre-theoretical understanding in which we already live. That tapping takes the form of an anticipatory projection of Dasein's Being (and of Being in general) which resembles nothing so much as Husserl's notion of *Vorzeichnung*, while the task of laying this out in an explicit ontology resembles nothing so much as Husserl's "explicative" analysis.

Heidegger of course is aware that "...we have no right to resort to dogmatic constructions and to apply just any idea of Being and actuality to this entity, no matter how 'self-evident' that idea may be...in itself and from itself." (SZ, § 5, 16/37) That is why he has recourse to the hermeneutic fore-structures--fore-having (*Vor-habe*), fore-sight (*Vor-sicht*) and fore-conception (*Vor-griff*)--which together make up the "hermeneutic situation" (§ 45, 232/275). This three-fold fore-structure is introduced as a way of securing our hermeneutical bearings, of seeing to it that our projective understanding is not capricious and free-floating. If the hermeneutic situation is secured, we can be assured that we have the entity as a whole in our grasp (fore-having), that we have sighted it correctly as to its mode of Being (*Seinsart*, fore-sight), and that we have a suitably articulate table of categories or conceptual framework within which to conduct our discourse (fore-conception). As a unit these three seem to me to amount to a spelling out and detailing of what Husserl called the *Vorzeichnung* which brings an entity into view, or of the horizon within which it appears. And like Husserl, Heidegger regards the fore-structures as revisable--they may be either final or provisional (*vorbehaltlich*) (SZ, § 32, 150/191) so that in Heidegger, Husserl and Gadamer there is a certain back and forth movement between what Heidegger calls understanding and interpretation until the right fit for the fore-structures is found, or until the *Vorzeichnung* is appropriately sketched.

But if Heidegger wants to avoid the wanton positing of just any projective framework at all, he also wants to insist that the fore-structures are not merely something to be tolerated, unavoidable limitations which ideally we would do without. On the contrary, the fore-structures are phenomenologically requisite, lacking which we would face the situation which Husserl describes in *Ideas I* as a material counter-sense, that is, an entity for which we have no preparatory predelineation and hence, about which we would lack even so much as a clue. The issue is hardly then to clear away all the fore-structures and to secure objective presuppositionlessness, but rather to see to it that our projective fore-structures are fitted and suited to gain us access to the things themselves:

What is decisive is not to get out of the circle but to come into it in the right way....In the circle is hidden a positive possibility of the most primordial kind of knowing. To be sure, we genuinely take hold of this possibility only when, in our interpretation, we have understood that our first, last and constant task is never to allow our fore-having, fore-sight and fore-conception to be presented to us by fancies and popular conceptions, but rather to make the scientific

theme secure by working out these fore-structures in terms
of the things themselves. (SZ, §32, 153/195)

By the time Heidegger reaches § 63 in *Being and Time* the
full hermeneutic situation for the interpretation of Dasein has
been secured. We have Dasein as a totality--from birth to death--
in our grasp (fore-having). We have gained access to Dasein in
its authenticity, and so we have both an adequate guiding concept of Dasein (fore-sight) and a table of existentialia with which
to articulate that concept (fore-conception). Hence Heidegger is
prepared to argue in § 65 that the Being of Dasein is temporality.
At that point the main argument of the existential analytic is consummated and the hermeneutic strategy of *Being and Time* vali-
dated. At that point he can claim that the projection of the Being
of Dasein in terms of existence has met with success and that he
has been able to let Dasein "put itself into words for the very
first time" (SZ, §63, 314/362). At this point we ought to be able
to recognize ourselves in the account which has been given of
Dasein in *Being and Time*. The work of phenomenological interpretation (*Auslegung*), of laying out an understanding in which
we already stand by way of projecting the Being of Dasein in the
appropriate way, is brought to a close.

HEIDEGGER AND HUSSERL

Let us now return to the question of the relationship
between Husserl's "pure" and Heidegger's "hermeneutic" phenomenology. It should be obvious that it is too facile to say that
Husserl's is a phenomenology without presuppositions while Heidegger's is a phenomenology in search of the right presuppositions. For whatever truth there is to this familiar claim, it is
perfectly clear that Husserl's theory of intentionality turns on a
theory of anticipatory predelineation. Far from differing on this
point, Heideggerian and Husserlian phenomenology are in fact in
agreement that perception or understanding must always be
guided beforehand by a preparatory grasp of what is to be
understood, failing which understanding is left disoriented and
cut adrift, unable to seize the matter at hand. Were consciousness
unable to sketch out beforehand the main lines of what is to be
perceived, then perception would break down and it would be left
with a chaos of uninterpretable, unintended hyletic data. Were
Dasein not equipped with a projective understanding of the entity
to be interpreted, it too would be left adrift and would need to

have recourse to the most commonplace and publically available framework of understanding, lest it understand nothing at all. That is why Husserl too can make his own the following text from *Being and Time*, if we read "presuppositions" in the sense of anticipatory predelineation:

> Philosophy will never seek to deny its presuppositions but neither may it simply admit them. It conceives them, and it unfolds with more and more penetration both the presuppositions themselves and that for which they are presuppositions. (SZ, §62, 310/358)

And that too is why the work of phenomenology for both Husserl and Heidegger is explicative: unfolding, explicating, laying out the implicit horizons which make explicit experience possible. Phenomenology for both is the *subtilitas explicandi*. For both phenomenological work lies in unfolding the horizons within which entities or objects appear. Derrida was quite right to say that in Husserl and Heidegger everything turns on the logic of implicit and explicit, which is why there is nothing vicious in the circular logic of phenomenology.

But if all this is so, then how are we to draw the lines of difference between Husserl and Heidegger, and what has become of Husserl's repeated insistence on presuppositionlessness? If up to now I have argued against this distinction it is because I take it to be, as it stands, too facile and global a characterization. I am prepared to reinstate it, however, in a more restricted and qualified sense.

For in Heidegger the projective character of the understanding is itself rooted in the more profound ontological make-up of Dasein; it reflects the Being of Dasein as care, existence and temporality (an ontology which, despite the protests of the Heideggerians to the contrary, Heidegger largely inherited from Kierkegaard.) Whence the hermeneutic circle was itself rooted in the Being of Dasein as care:

> What common sense wishes to eliminate in avoiding the circle, on the supposition that it is measuring up to the loftiest rigour of scientific investigation, is nothing less than the basic structure of care. Because it is primordially constituted by care, any Dasein is already ahead of itself. As being, it has in every case already projected itself upon definite possibilities of its existence; and in such existentiell projections

has, in a pre-ontological manner, also projected something like existence and Being. *Like all research*, the research which wants to develop and conceptualize that kind of Being which belongs to existence is itself a kind of Being which disclosive Dasein possesses; can such research be denied this projecting which is essential to Dasein? (SZ, §63, 315/363)

The projective character of understanding for Heidegger is a function of the Being of Dasein as a project. The understanding of Dasein proceeds by way of a projective predelineation because the Being of Dasein is being-ahead-of-itself. We said above that Heidegger's conception of hermeneutics went hand in hand with an ontology of Dasein as existence. And we pointed out that for Heidegger hermeneutics was the counter-part, the reverse movement, of an existential drift towards fallenness, and hence that for Heidegger hermeneutic projection always required a certain violence, that is, an effort to resist the drift and pull of the commonplace presuppositions of average everydayness.

Now because the phenomenological structure of projection is situated within the ontological structure of care, it follows first of all that the very notion of presuppositionlessness is an ontological paradox for Heidegger. To acquire presuppositionlessness Dasein would *per impossibile* have to undergo a change in Being so that it would no longer be Dasein but something else which is neither futural nor projective. It furthermore follows, according to the basic meaning of care, that Dasein is thrust or thrown into its presuppositions, for care is thrown as well as projective, factical as well as existential. Hence Dasein always moves about within a certain historically situated, factical understanding, or rather pre-understanding, from which it neither can nor wants to extricate itself. And that is the point of departure for Gadamer's extension of *Being and Time* in the direction of a "philosophical hermeneutics."[12]

In the case of Husserl, however, this Kierkegaardian existential ontology of care is nowhere to be found. Husserl makes no effort to root the analysis of predelineation in an ontological subsoil. It is his intention that the theory of *Vorzeichnung*, indeed the entirety of phenomenological theory, be ontologically innocent and unencumbered with ontological presuppositions. Phenomenology ought to have a more straightforward descriptive sense; it ought to arise only from the laborious work of phenomenological reflection and have no part in a larger ontological program. Husserl thinks that the reflective life of the phenomenological ego can make its way without ontological guidance, and hence that the discovery of intentional predelineation is not itself ontologically predelineated but is ontologically neutral.

Now at last the issue between Heidegger and Husserl is out in the open. It has to do with the question of *ontological* presuppositions, not with the concrete life of the worldly ego (=Being-in-the-world). I do not see that there is any serious difference between Heidegger and Husserl on the question of the projective-ness of intentional life, of its dependence upon anticipatory structures in order to make its way around the world. The real issue arises on the level of whether or not this phenomenological account harbors within it an ontological view. And of course Husserl contends that it need not and does not.

But Heidegger questions whether in fact Husserl succeeds in this, or whether the commitment to ontological presuppositionless-ness does not itself conceal within it a certain ontology, a certain way of understanding the Being of consciousness, the *Sein* of *Bewusstsein*. Heidegger thinks that at this point Husserl wants to walk on water. He thinks that every attempt to proceed without ontological bias ends up being subverted from behind by an onto-logical presupposition which is at work on it behind its back. Thus for Heidegger the very *attempt* to proceed without ontologi-cal guidance is itself inspired by a certain ontology, viz., a Cartesian ontology which presupposes the separability of reflec-tive consciousness from concrete first-order experience which is embedded in language, historical tradition and culture. Husserl does not deny--in fact he clearly explicates the fact--that first-order intentional acts are inhabited by a ring of presuppositions which make them possible. That is his express teaching. His dif-ference with Heidegger has to do with the capacity of conscious-ness to effect a second-order, reflective extrication from such conditioned and predelineated understanding. Husserl wants to hold that reflective consciousness is conducted under conditions which he precisely denies of first-order experience. Husserl clings to the ideal that the reflective ego enjoys a mode of inten-tional life--free from potential, implicit, horizonal and predelinea-tory factors--which he otherwise insists belong to the very make-up of intentional life at large. For Heidegger this is an inconsis-tency which arises from Husserl's ontological presuppositions, that is, from his acceptance of the possibility of pure reflection, of the ideal of transcendental consciousness, of the separability of reflecting consciousness from first-order experience. That convic-tion is laden with ontological deposits and belongs to the mainstream of Western metaphysics from Plato to Hegel.

The point which the Heideggerian would make here can also be put as follows. Whenever Husserl undertook to describe the *concrete* functioning of intentional life, he invariably resorted to a hermeneutic schema; whenever he *actually practiced* the phenomenological method, he ascribed to consciousness a hermeneutic, contextualized composition. The life of the living ego was for him always ringed about by a border of horizonal structures, always temporally situated within a stream of protended and retained moments, laced with potential and implicit structures, all of which make it possible to "interpret" (*auffassen*) the intentional object. The notion of a presuppositionless grasping of the intentional object, of grasping it without the support of these horizonal structures, is unheard of on this level. The idea of presuppositionlessness arises only when Husserl wants to characterize the nature of phenomenology *as a science*. At this point he has recourse to the inherited idea of a Cartesian science, of an absolutely presuppositionless beginning. This idea does not arise from his concrete phenomenological inquiry. Rather it is invoked as a way of settling the ongoing debate about psychologism, naturalism and historicism. But from the Heideggerian point of view this Cartesian ideal has no phenomenological credentials. For the Heideggerian, phenomenology not only has a hermeneutic component; it is *inherently* hermeneutic, whereas the goal of presuppositionlessness is a Cartesian import, a residue of the modern metaphysical tradition which from Descartes on is focused on the debate over the idea of transcendental consciousness.

And so it seems to me that one does in the *end* come back to the opposition of Husserl's presuppositionless phenomenology with Heidegger's hermeneutic phenomenology. But one should not *begin* there. Rather one should first examine the extensive preparations which are made in Husserl's epoch-making investigations in *Ideas I* which opened up the horizon within which hermeneutic phenomenology appeared a little over a decade later.

NOTES

1. SZ = Martin Heidegger, *Sein und Zeit*, 10. Aufl. (Tübingen: Niemeyer, 1963). The page numbers following the slash are to the English translation: *Being and Time*, trans. John Macquarrie and Edward Robinson (New York: Harper & Row, 1962).

2. Edmund Husserl, "Philosophy as a Strict Science," in

Phenomenology and the Crisis of Philosophy, trans. Q. Lauer (New York: Harper Torchbooks, 1965), p. 146.

3. CM = Edmund Husserl, *Cartesianische Meditationen* in *Husserliana*, Band I, ed. S. Strasser (The Hague: Nijhof, 1973); the page numbers following the slash are to the English translation: *Cartesian Meditations*, trans. Dorian Cairns (The Hague: Nijhoff, 1960).

4. Edmund Husserl, *Analysen zur Passiven Synthesis*, *Husserliana*, B. XI, ed. M. Fleischer (The Hague: M. Nijhoff, 1966), p. 11.29-32.

5. This point is made in S. Stephen Hilmy, "The Scope of Husserl's Notion of Horizon," *The Modern Schoolman*, 59 (November, 1981), 21-48, that the notion of horizon applies not only to perception but to intentional life in general. For further investigations into the notion of horizon, see: Helmut Kuhn, "The Phenomenological Concept of 'Horizon'," in *Philosophical Essays in Memory of Edmund Husserl*, ed. Marvin Farber (Cambridge: Harvard Univ. Press, 1940), pp. 106-23; Cornelius Van Peursen, "The Horizon," in *Husserl: Expositions and Appraisals*, ed. F. Elliston and P. McCormick (Notre Dame: University Press, 1977), pp. 182-201; Henry Pietersma, "The Concept of Horizon," *Analecta Husserliana*, 2 (1972), 278-82.

6. Paul Ricoeur has pointed to *Auffassung* in the *Logical Investigations* as a hermeneutic element in Husserlian phenomenology; he also discusses the notion of *Auslegung*, but with results which differ considerably from ours. Ricoeur does not discuss the central role of *Vorzeichnung*. See his "Phenomenology and Hermeneutics," in *Hermeneutics and the Human Sciences*, ed. & trans. John Thompson (Cambridge: University Press, 1981), pp. 101-28.

7. Edmund Husserl, *Erste Philosophie* (1923/24), II, *Husserliana*, VIII, ed. R. Boehm (The Hague: Nijhoff, 1959), p. 148.1-8.

8. Husserl insists that this process of *Erfüllung* is never finished:

This sense, the *cogitatum qua cogitatum*, is never present to actual consciousness as a finished datum; it becomes "clarified" only through explication of the given horizon and the new horizons continually awakened. (CM, §19, 82-83/45)

9. Ricoeur makes this point quite well in his commentary on Husserl's *Ideas I*. See the translator's notes to §§ 47-9 in *Idées directrices pour une phénoménologie* (Paris: Gallimard, 1950).

10. All references to *Ideas I* are to Edmund Husserl, *Husserliana*, B. III, 1, *Ideen zu einer reinen Phänomenologie und phänomenologische Philosophie*. Erstes Buch, *Allgemeine Einführung in die reine Phänomenologie*, ed. Karl Schuhmann, (The Hague: Nijhoff, 1976). The page citations are to the marginal pagination of the 1916 edition, preserved in the Husserliana editions, and then, following the slash, to the English translation: *Ideas Pertaining to a Pure Phenomenology and to a Phenomenological Philosophy*, First Book, *General Introduction to Pure Phenomenology*, trans. Fred Kersten (The Hague: Nijhoff, 1982).

11. Jacques Derrida, *Margins of Philosophy*, trans. Alan Bass (Chicago: University Press, 1982), p. 126.

12. Hans-Georg Gadamer, *Wahrheit und Methode: Grundzüge einer philosophischen Hermeneutik*, 4. Aufl. (Tübingen: Mohr, 1975), especially pp. 240-56; *Truth and Method* (New York: The Seabury Press, 1975), pp. 225-240.

7. SAMUEL IJSSELING

*HEIDEGGER AND THE DESTRUCTION OF ONTOLOGY**

In § 5 of *Die Grundprobleme der Phänomenologie*, Heidegger discusses the methodological character of ontology. According to this text, the method of ontology is phenomenology, phenomenology itself being characterized by *three basic components (drei Grundstücke):* the phenomenological reduction, the phenomenological construction and the phenomenological destruction. Heidegger remarks:"In accordance with their very content, reduction, construction and destruction belong together and must be founded in their belonging together" (GP, p. 31/23).[1]

Ontology, and in this context ontology is for Heidegger the same as philosophy, poses the question about Being. "Being is the genuine theme, the only theme of philosophy" and philosophy is "the theoretical-conceptual interpretation of Being, its structure and its possibilities" (GP, p. 15/11). Being is also always the Being of a being, the Being of something that is, and the question about Being must therefore in the first instance and necessarily take its departure from a being. Being is always Being that is understood. "There is Being, only when the understanding of Being, that is, *Dasein*, exists" (GP, p. 26/19). The understanding of Being, however, is "not a kind of cognition. Rather, it is the basic determination of existing" (GP, p. 393/278). It belongs to the very structure of *Dasein*, that is to say, to the very possibility that there be something to see, something to do and something to say. Without a (preontological) understanding of Being, not a single being would be accessible to us. "No comportment toward beings could exist, which did not understand Being" (GP, p. 466/327). And "only in the light of

---

*From Samuel IJsseling, "Heidegger and the Destruction of Ontology, *Man and World*, 15(1982), pp. 3-16. Reprinted by permission of Martinus Nijhoff, Publishers B.V.

the understanding of Being can we encounter beings as beings"
(GP, p. 390/275). Neither, however, is an understanding of Being possible apart from a relation to beings. "No understanding of
Being would be possible, which were not rooted in a certain comportment toward beings" (GP, p. 466/327). Furthermore, Being is
always understood in a certain light and against an historically
changing horizon of understanding. This light and this horizon
form both the conditions for the possibility and the limitation of
the understanding of Being. Ontology, therefore, asks about the
basic structures and various modifications of the understanding of
Being. It also asks about the light and the horizon which found
and delimit this understanding. "What is it that makes this understanding of Being possible?" "Out of what pregiven horizon do we
understand something like Being?" (GP, p. 21/16). Thus runs the
central question of Heidegger's thinking.

With this, the *three components of phenomenology* are given in
principle:

1) "Ontological investigation concerns itself initially and
necessarily with beings, but is then in a certain manner led away
from beings and back to their Being" (GP, pp. 28-29/21). Heidegger calls this *leading back from beings to Being* the phenomenological reduction. In doing so, he remarks that he is employing
a central term in the phenomenology of Husserl, but only "in
respect of the word and not of the thing". Indeed, there is a
fundamental and radical difference between the Heideggerian and
the Husserlian reduction. Heidegger formulates this difference as
follows:

For Husserl, the phenomenological reduction is the method of
leading phenomenological viewing from the natural attitude of
human involvement in the world of things and persons back to
the life of transcendental consciousness and its noetic-
noematic processes of experience in which objects are constituted as correlates of consciousness. *For us*, the phenomenological reduction means leading phenomenological viewing from
the always differently determined conception of beings back to
the understanding of the Being of these beings, back, that
is, to the projection of the mode of the unconcealedness of
the Being of these beings. (GP, p. 29/21).

In other words, for Husserl, *the wonder of wonders* was transcendental subjectivity, "behind which it would be nonsensical to want to investigate further";[2] for Heidegger, the *wonder of wonders* is that beings are (Nachwort, WiM, pp. 46-47/355).

2) To be sure, Being and beings are "unveiled equiprimordially" (GP, p. 456/320), but "Being is not accessible in the same way as beings. We do not simply find it lying before us. Rather, it must in every case be brought into view by way of a free projection *(Entwurf)*". As Heidegger says, "we designate this projection of pregiven beings against the background of their Being and the structures of their Being as phenomenological construction" (GP, pp. 29-30/22). In saying this, we should observe further that every great philosophy is characterized by just such a free projection. As such, every great philosophy is a construction and contains its own *'thesis concerning Being' (These über das Sein)*.

3) "Because in accordance with its own existence *Dasein* is historical, the possibilities for gaining access to beings and the modes for the interpretation of beings are themselves variable and do in fact vary in different historical situations." Every ontology, including that of Heidegger himself, "is determined by its historical situation and thus at once by certain possibilities for gaining access to beings as well as by the preceding philosophical tradition." For this reason, there

> ...necessarily belongs to the conceptual interpretation of Being and its structures, that is, to the reductive construction of Being, a *destruction*, that is, a critical dismantling of the concepts which have been handed down to us and which we initially have to employ, a dismantling which proceeds to the sources out of which such concepts have been drawn. Only by way of the destruction can ontology provide itself with full phenomenological assurance of the genuineness of its concepts. (GP, p. 31/22-23)

This phenomenological destruction can perhaps be brought into connection with the Husserlian *epoche*, at least with the *epoche* of the later Husserl, where it is no longer a question of "bracketing the world" but rather of *bracketing* the various interpretations of the world. Here, however, there is once again a great difference between Husserl and Heidegger. For Heidegger, it is not a question of placing the traditional ontology between brackets and still less of annihilating this ontology. Heidegger writes the following: "The destruction does not seek to bury the past in nothingness.

It has, rather, a positive intention." The destruction is "neither a negation of the tradition nor a condemnation of the tradition to nothingness. On the contrary, it is precisely a positive appropriation of this tradition" (GP, p. 31/23). In the *destruction*, it is a question of freeing oneself *from* the traditional stock of ontological concepts and words which dominate and determine both our thinking and the fulfillment of our humanity--and which do so all the more to the extent that we fail to become aware of and recognize their domination and determination. At the same time, however, it is a question of freeing oneself *for* "the original experiences in which the first and thenceforth leading determinations of Being were gained" (SZ, p. 22/44). As we shall see, the *destruction* is ultimately oriented so as, by means of a structural analysis of factically existing philosophy, to bring attention to precisely what it is that happens whenever a philosophy is built up and formulated, whenever it comes about and is realized. In *What is Philosophy?* one can read the following:

> Destruction does not mean demolishing or destroying. Rather, it means dismantling, that is, pulling down, taking apart and setting aside--namely, the merely historical assertions concerning the history of philosophy. Destruction means: opening our ears, freeing ourselves to hear what addresses us in the philosophical tradition as the Being of beings. (WP, pp. 33-34/71-73)

Although Heidegger has on several occasions expressed himself with clarity concerning the meaning of the *destruction*, the entire project nevertheless remains somewhat ambiguous. This ambiguity shows itself, as well, in the very term 'destruction' which literally does mean something on the order of destroying. But, in fact, its intention is to be a structural analysis of the tradition. The ambiguity of the entire problematic is also evident from the still prevalent misunderstanding regarding the destruction, which one encounters in most of the interpreters of Heidegger's thinking. It is precisely this situation which has led Derrida, in my view rightly, to prefer the term 'deconstruction'.

In what follows, we shall try to gain some insight into the basic structure of the *destruction*. It is extremely important that we do have insight into this structure, for (1) according to Heidegger, "it is only in the carrying out of the destruction of the ontological tradition that the Being-question attains its true concretion" (SZ, 26/49). What precisely the *concretion (Konkretion)*

consists in must, of course, be clarified. Ultimately, however, as
we shall see, the *destruction* and the question about Being coincide to a certain extent. (2) The seed of Heidegger's break with
(Husserlian) phenomenology lies in the problematic of the *destruction*. For with Heidegger, the meaning of *die Sache selbst* alters
and comes more and more to lie in thinking and what has been
thought, in speaking and what has been said, in philosophy as a
work, a fact, and so on.

Let us now elaborate somewhat further the problematic of the
destruction. We shall do so, in the first place, against the background of *fundamental ontology (Fundamentalontologie)*; secondly,
in connection with the problematic of the reduction and the construction and in connection with the *retrieval (Wiederholung)* of
the Being-question; and finally, in the light of the later work of
Heidegger where the problematic of the destruction *(Destruktion)*
returns in the form of the *overcoming of metaphysics (Ueberwindung der Metaphysik)*, the *return into the ground of metaphysics
(Rückgang in den Grund der Metaphysik)* (which is, in fact, the
title of the later attached foreword to *What is Metaphysics?*), and
the *step back (Schritt zurück)* "out of metaphysics into the
issuant and abiding essence *(Wesen)* of metaphysics," "out of
technology into the issuant and abiding essence of technology,"
or "out of the already thought into an unthought whence what
has been thought receives room to issue and abide in its essence"
(ID, p. 44/48).

I. FUNDAMENTAL ONTOLOGY

In *Being and Time*, Heidegger formulates the question about Being in the form of a *fundamental ontology*. The purpose of this
fundamental ontology is twofold and, from the very beginning,
also somewhat ambiguous. On the one hand, Heidegger wants to
lay (or perhaps lay bare) the foundation for a possible ontology,
that is to say, for a meaningful speech concerning beings as beings and concerning the Being of beings. On the other hand, he
wants also to provide a fundamental explanation of the already
existing ontologies, from Aristotle right up to the present day,
not only the factically and explicitly developed ontologies as one
encounters them in the great philosophers, but as well the rather
more implicit ontologies which are at work in the sciences and in
our practical involvement with beings. This fundamental clarification attempts to recover the structure and the origin, the

conditions for the possibility and the consequences, the birth certificates and the genealogy of every actual and possible ontology or of every actual and possible *thesis concerning Being*. What is at issue is a *fundamental ontology* "out of which," as Heidegger expressly says, "all other ontologies can arise" (SZ, p. 13/34).

We should not forget that Heidegger's thinking seeks to carry out a *retrieval* of the project of Kant. Kant had attempted, on the one hand, to lay the foundation for a possible and accountable metaphysics. He wrote the *prolegomena* "to any future metaphysics which will be able to arise as a science." Kant finds the foundation in the 'I think' (the subject or consciousness) which is essentially dependent upon and referring to experience. On the other hand, Kant also tried to provide a radical explanation of the factical and historically existing metaphysics. According to Kant, these metaphysics are an effect of the *transcendental illusion*, that is, the ineradicable inclination of mankind to think and to reason without testing its thoughts and reasonings anew against experience. Man then constructs all manner of systems and builds elaborated conceptual palaces which are no doubt impressive and attractive but are in fact illusory. Kant will say that such illusions are never wholly to be avoided, because human beings actually have need of illusions in order to be able to live. The illusions can, however, be made harmless. Heidegger will look elsewhere than Kant for the foundation of a possible ontology and the fundamental explanation of all factically existing ontologies. He seeks these in the structure of *Dasein*, understood as essentially *discovering (entdeckend)* and *covering over (verdeckend)*. Nevertheless, he stands squarely within the post-Kantian tradition in which it is part of the task of every philosophical speech to account radically for its own speech and radically to explain all factically existing speech.

In order to build a *fundamental ontology*, therefore, two things are necessary. As Heidegger states it, there is "a twofold task involved in the elaboration of *fundamental ontology*" (SZ, p. 15/36). A structural analysis of *Dasein (Analytik des Daseins)* must be given on the one hand and, on the other hand, a structural analysis of the already existing ontology *(Destruktion der bisherigen Ontologie)*. Both the *analysis of Dasein* and the *destruction* are somewhat ambiguous. This ambiguity will later bring Heidegger to formulate the Being-question in a rather different way.

On the one hand, the analysis of *Dasein* is an analysis of

human being conceived as *Dasein*. On the other hand, it is an analysis of *Being* insofar as Being is *there*, of the *Sein* that is *Da*, that takes place whenever a human being fulfills its own Being. This analysis yields the so-called existentials. We need not discuss their problematic again here for it is familiar enough. Only the following must be recalled. *Dasein* is structurally characterized by a fundamental openness to the world and to itself *(disclosedness, Erschlossenheit)*. This openness, that is to say, the possibility that there be something to do, to see and to say, is of a primarily practical nature. Man is in a position to be involved in a more or less intelligent and understanding way with himself, his fellow men and the things of his environment *(understanding) (Verstehen)*. Man can also involve himself theoretically with himself and his environment. He can observe and study himself, others and the world. He can develop a science and design theories. This theoretical attitude, however, is always secondary. It is merely one determinate way of more or less intelligent and understanding involvement with oneself and one's world. Such intelligent and understanding involvement in or commerce with a world also implies an *understanding of Being (Seinsverständnis)*, an understanding which is primarily *preontological (vorontologisch)* but which can be ontologically explicated. Without this *understanding of Being*, there would be nothing to do, to see or to say. "No comportment toward beings could exist, which did not understand Being" (GP, p. 466/327). And "only in the light of the understanding of Being can we encounter beings as beings" (GP, p. 390/275).

Understanding exhibits a determinate structure. A determinate facticity is interpreted and taken up in the light of a future *(Zukunft)* which comes toward us and is anticipated by us. The whole of human existence consists from moment to moment in a continuous, interpretive taking up of oneself and one's world, and a continuous, anticipative breaking open or holding open of a future. Human existence, therefore, presupposes time as a condition for its possibility.

Furthermore, understanding has a more concrete and better articulated existence in language, particularly in texts or works (whether literary or philosophical), and so in the whole of human institutions, social institutions, human attitudes and modes of human behavior and comportment (for example, technology). Language and word, institution and attitude, all belong to *Dasein*; they are *daseinsmässig*. Through and in all this, the human being fulfills his own Being and Being itself happens, that is, takes place. According to Heidegger, Being does not happen somewhere in the heavens or among the stars. Rather, it takes place precisely there, where man fulfills his humanity, his being-human. As the *'there'* of *Being*, the *Da des Seins, Dasein* is the place where Being takes place.

Finally, understanding can be both authentic and inauthentic. It is inauthentic whenever it does not spring from and return to a primordial experience of and a genuine proximity to itself and its world. Such experience and proximity can be ruined by and buried under all manner of traditional formulas and words, concepts and theoretical constructions, existing attitudes and works. These may well have come about--though usually only partially--within a primordial seeing, but in the course of history they have taken on a life of their own and disengaged themselves from experience and proximity. They thus come to be adopted and repeated uncritically, without it being realized exactly what is happening here. Moreover, many concepts and words stemming from a particular domain of experience have become mingled with concepts and words from an entirely different domain of experience. Thus, according to Heidegger, many concepts belonging originally to the world of manufacture (production) have been taken over into areas where there is no talk at all of manufacture. Similarly, many words stemming from the typically Greek world have been 'translated' into Latin and taken over into the entirely different Roman (Christian) world. Finally, there are many words which functioned originally in a religious and theological context but which have slipped over into philosophical speech. None of this, of course, is without consequences. The primordial proximity to oneself and the world has been lost. This is the case, among others, whenever man interprets and conceives himself as a being that is present in the midst of other beings, whenever he interprets and conceives the world as a totality of beings spread out before him as a sort of spectacle and whenever he understands *Being (Sein)* as a being *(Seiendes)*, as "permanent presence" *("stdndige Anwesenheit")* (EM, p. 154/169), or in an ontotheological framework.

One may not, however, think too simplistically about primordial experience and genuine proximity, for human Being is by its very nature interpretative and it is always already interpreted Being just as the world is always an already interpreted world. Outside the interpretative taking up and projecting of oneself and the world, there can be no talk of either a meaningful existence or a meaningful world. Primordial experience and genuine proximity must therefore be conceived as an experience of and a proximity to this interpretative taking up and projecting, or as an experience of and a proximity to the very fulfilling of human Being. It is not a proximity to present beings but a proximity to the happening that takes place whenever human Being fulfills its existence. In the *Prolegomena zur Geschichte des Zeitbegriffs* (lectures from 1925) Heidegger writes the following:

> It is simply a matter of fact..., that our most straightforward perceptions and conceptions are already expressed and, what is more, in a certain way already interpreted. We do not primarily and originally see objects and things; rather, we initially speak about them. More precisely, we do not say what we see regarding objects and things; on the contrary, we see in such things merely what one commonly says about them. This characteristic determinateness of the world and the possibility of apprehending and comprehending the world in virtue of this expressness, in virtue of certain things already having been said and discussed, is...precisely what must be brought fundamentally into view. (PGZ, p. 75)

The text just cited deals primarily with the problematic of *categorial intuition (kategoriale Anschauung)*, a problematic of Husserl which was of the greatest importance for Heidegger in formulating the Being-question. Unfortunately, we cannot go into that matter here. It is, however, important to understand that it is Heidegger's concern to think the fulfillment of human Being, *Dasein* as *speaking* and to inquire about the Being *(Sein)* of that distinctive being *(Seiendes)* "which is speaking itself"(PGZ, p. 203). Phenomenology is, therefore, "neither a philosophy of intuition nor a philosophy of the immediate" as Rickert had supposed. According to Heidegger, Rickert had utterly failed to grasp the meaning of phenomenology. "In the demand for an ultimate and direct givenness of the phenomena there is implied no such comfort as that of an immediate viewing" (PGZ, pp. 120-121).

On the basis of the foregoing, it should be clear what the *destruction* is at the level of fundamental ontology. As formulated in the *Prolegomena*, destruction is "a letting-see which lays bare and which is to be taken in the sense of a methodically conducted dismantling of all (accidental and necessary) coverings-over *(Verdeckungen)*" (PGZ, pp. 118-19). In the *Grundprobleme*, we read the following description of the destruction: Destruction is "a critical dismantling of the concepts which have been handed down to us and which we initially have to employ, a dismantling which proceeds to the sources out of which such concepts have been drawn" (GP, p. 31/22-23). It is a question of freeing ourselves from something, which--at least initially--we can neither do without nor get outside of. This liberation cannot come about by denying or refuting the factically existing philosophies. Still less can it occur by acting as if no one had ever philosophized before. There is no philosophical nullpoint and it is simply impossible to begin all over again. It is even the case, on the one hand, that the already existing philosophies weigh upon us as a burden and dominate our thinking--all the more so in the degree that we fail to become aware of and recognize this domination as such. On the other hand, however, it is precisely the philosophical tradition which makes our own thinking possible. Without a tradition there could be no philosophy. The factically existing philosophies form a part of our reality. As philosophies, they belong to *Dasein*. But the emancipation of philosophy can only consist in tracking down the way in which philosophy is even possible and first comes about, what is revealed and concealed in and by philosophy, what is excluded and forgotten in and by philosophy, what displacements and alterations have arisen within philosophy, how philosophy functions in our world and what it accomplishes and has accomplished. One must ask oneself what precisely happens whenever philosophy or *a* philosophy is constituted, makes an appearance, becomes public, is built up and expanded, comes to completion and eventually goes to its end. The basic question of Heideggerian thinking runs as follows: What precisely takes place when philosophizing occurs? More concretely, Heidegger says, for example, near the end of his book on Kant and in regard to his interpretation of Kant: Our task is "to inquire not about what Kant says but about what it is that *happens* in the course of his attempt to lay a foundation (for metaphysics). The foregoing interpretation of the *Critique of Pure Reason* aims exclusively at laying bare this *happening*" (KPM, p. 193/221). He says nearly the same thing regarding his interpretations of Schelling and Hegel, Leibniz and Descartes. Stated more generally, Heidegger poses questions such as "What is metaphysics?" and "What is philosophy?" in which the *is* must be understood transitively, just as he himself remarks (WP, p. 22/49). What is it that happens in philosophy? What makes philosophy be what it is and be such as it is? Or, as Heidegger puts it, "*Was heisst Denken?*" where *heissen* means something like 'command', 'call up', 'summon into existence', 'provide direction', and so on. For

Heidegger, asking what precisely happens when philosophizing occurs, coincides with posing the question about Being, the question about the Being of that distinctive being which is philosophy. The *destruction of ontology* and the Being-question belong to each other. Let us attempt to elucidate this in another manner.

II. *THE QUESTION ABOUT BEING*

Being is always the Being of being. The question about the Being of beings "can be asked in respect of *every* being" (PGZ, p. 186). Nevertheless it is not a matter of indifference which being one selects as a model or as a point of departure. According to Heidegger, it is striking that in asking the question about Being the entire tradition tends to take its departure from things which one can represent to oneself and observe, from objects which in one manner or another have been manufactured (produced) or possibly even from a highest being "in which", as one is perhaps inclined to support, "the idea of Being in general is realized in the most genuine sense" (PGZ, p. 233). The choice of such exemplars is not without consequences for the answer to the question about Being or, better stated, with this choice it is presupposed that the question has already been answered. Therefore Heidegger takes another example as a model or as a point of departure for the Being-question. In the very first paragraphs of *Being and Time*, Heidegger speaks of the "ontic and ontological priority of the Being-question". Though it is usually forgotten, this has a twofold meaning. It means that in philosophy the question about Being deserves priority above all other questions: "Being is the genuine theme, the only theme of philosophy" (GP, p. 15/11). It also means, however, that *as a question*, the question about Being is the being *par excellence* which must serve as the point of departure for the question about the Being of beings. In the *Prolegomena*, Heidegger writes the following: "questioning itself is a being *(Seiendes)*" (PGZ, p. 199). What must be done? We must "lay bare questioning as a being, that is, we must lay bare *Dasein* itself" (PGZ, p. 200). Stated differently, our task is to ask about the Being *(Sein)* of that distinctive being *(Seiendes)* "which *is* speaking itself" (PGZ, p. 203). Speaking, and therefore also questioning and answering, is "just as *Dasein* is, that is, it exists" (GP, p. 296/208). It was from Husserl that Heidegger first learned to regard a question (and also an answer) as a being, and it was Heidegger's view that this way of regarding questions and answers belongs to the essence of phenomenology.

Husserl, however, neglected to inquire about the Being of such a being (PGZ, p. 157). This was, furthermore, no "mere omission or oversight of a question which should have been posed" (PGZ, p. 178). Rather, it hangs together with "the history of *Dasein* itself" or with "the kind of happening characteristic of *Dasein*" (PGZ, p. 179).

For Heidegger, the concern is to approach philosophy as a network of questions and answers, as a *work of language*, as a being, and then to ask oneself what the *Being (Sein)* of this work (being, *Seiendes*) is, what precisely happens in and through this work. The approach to philosophy as a work, and above all as a *work of language*, occurs most explicitly in *The Origin of the Work of Art*. This essay deals not only with art as a work but also with philosophy as a work. In addition to the work of art, the following are named as works: "the deed which founds a state", "the nearness of that, which is not simply a being, but is rather the uttermost being of all beings *(das Seiendste des Seienden)*", "the essential sacrifice", and "the questioning that belongs to a thinking which, as the thinking of Being, names Being in its questionworthiness" (H, p. 50/62). Elsewhere, namely in the *Introduction to Metaphysics*, a text which was written at approximately the same time as *The Origin of the Work of Art*, Heidegger offers the following enumeration of works: "the work of the word as poetry, the work of stone in the temple and the statue, the work of the word as thinking, the work of the *polis* as the historical place which founds and preserves all this" (EM, p. 146/160). Heidegger adds to this list the following characterization: "In accordance with what was said earlier, 'work' is always to be understood here in the Greek sense as *ergon*, as the presence that has been set forth into unconcealedness." Immediately preceding the enumeration of different forms of work we find the following utterance: "Unconcealedness comes about only insofar as it is brought about, worked out of concealedness *(erwirkt)*, by the work." In accordance with the familiar words from *The Origin of the Work of Art*, it is said of every work, even of philosophy, that it is "the setting itself into work of truth." As a work, philosophy is a "work of truth". Heidegger remarks here that such expressions are distinguished by an essential ambiguity, because "truth is once a 'subject' and again an 'object'", albeit "both characterizations remain inappropriate" (H, p. 64/77). The work is the bringing about of unconcealedness *(aletheia)* and unconcealedness takes place only in and through the work. At the same time, the work is also brought

about, worked out into the open *(bewirkt)* by truth as unconcealedness. This holds in an exemplary fashion for philosophy as
a work or for the *thinking of Being*, where the genitive is just as
well a *genitivus objectivus* as a *genitivus subjectivus* and the
words 'subjective' and 'objective' are likewise inadequate.

To be sure, this is not the only thing that Heidegger says
concerning the work in *The Origin of the Work of Art*. It would
be well worth the effort to take up anew Heidegger's text on art
as a work and read it sentence for sentence as a text on *philosophy* as a work. In doing so, one should pay particular attention
to what Heidegger says in his later *Addendum*:

> The entire treatise 'The Origin of the Work of Art' moves
> knowingly, though not expressly, upon the path of the ques-
> tion concerning the issuant and abiding essence of Being.
> The attempt to situate the sense *(Besinnung)* of what *art*
> [philosophy] is, is determined entirely and decisively by the
> question about *Being*. (H, p. 73/86)

Such a rereading is, of course, not possible here. We can, however, make a few remarks in respect of the possibility of
approaching philosophy as a work, taking our lead from *The
Origin of the Work of Art*.

It can first of all be said negatively that one should avoid
three things: (1) viewing the work that philosophy is, purely as
a product of man; (2) viewing the work in the light of the conception of truth as adequation, that is to say, thinking that
philosophy is a more or less adequate rendition or representation
of a reality given outside philosophy; (3) viewing the work as a
sign or a network of signs which is supposed to signify a reality
given outside the work. Whenever one conceives and approaches
philosophy in this manner, one is already the victim of precisely
that ontology which Heidegger attempts to subject to a *destruction*, that is, one is a victim of what Heidegger calls metaphysical
thinking.

It can then be said positively that every great philosophy is
a structure, a work of building *(Bauwerk)*, a construction. As a
construction, it is not a depiction or representation of the world.
On the contrary, it founds and establishes a world. The constructed work of philosophy, a philosophical text stands there
just as the temple at Paestum stands there, and in standing there
it opens a world, offers a view to men and gods, and lets things
become visible. Philosophy is a place where reality comes at once

to be unveiled and veiled. It is precisely on the basis of this veiling and unveiling that there is something on the order of what we call world. The constructed work that is philosophy cannot exist without man, but it is not merely the product of man. Constructing a philosophy is, above all things, a matter of receiving and standing open, perceiving and listening. In a certain sense, philosophy constitutes itself. At the same time, philosophy is not a *creatio ex nihilo*. A philosophy is necessarily built up out of a pregiven material. This material is not the stone, the pigment, the color, as in architecture or painting. Rather, it is the words as in poetry. As Heidegger constantly emphasized, from the earliest *Marburg Lectures* to his latest publications, these words may not be understood as signs. The word is not a sign. It does not signify something which would be given or present somewhere outside and surrounding the word. As Heidegger says in *What Is Called Thinking?*, the issuant and abiding essence of saying does not allow itself to be determined on the basis of the sign-character of the word (WD, p. 123/202). Saying *(Sagen)* is showing *(Zeigen)*. Heidegger goes so far as to contend that the moment in which the word shifted from being *something that shows (Zeigendes)* to being a *sign (Zeichen)* was one of the most decisive moments in the history of truth, the history of the understanding of truth as agreement or correspondence and of Being as "permanent presence" *("ständige Anwesenheit")*. In *The Origin of the Work of Art*, we read the following:

> Where no language issues and abides *(west)*...there can be no openness of the being *(des Seienden)* and in consequence no openness of the not-being *(des Nichtseienden)* and of the void. When language names the being *(das Seiende)* for the first time, such naming first brings that being *(Seiendes)* to expression, that is, to the word and to appearance. This naming first nominates the being *(das Seiende)* to its Being *(Sein)* from its Being *(Sein)*. (H, p. 60/73)

Elsewhere, Heidegger says the following: "Things first come to be and are in the word, in language" (EM, p. 11/11). "Language first grants and warrants the possibility of standing in the midst of the openness of the beings *(des Seienden)*" (HD, p. 35) and "did our ek-sistent essence not stand in the power of language, then all beings *(Seiende)* would remain closed off to us: the being that we ourselves are, no less than the beings that we ourselves are not" (EM, p. 63/69). What is said here concerning language holds in exemplary fashion for the language of poets and thinkers and for the *work of language* that is philosophy and poetry. This, therefore, is the meaning of the constantly returning sentence in *The origin of the Work of Art*: "The work is the setting itself into work of truth." To the material out of which a philosophy is necessarily built up belong the fragments and passages which are and must be taken over from other works. No single work, no single text, ever stands wholly upon itself. It always refers to other texts upon which this one depends. A text is always taken up into a "context of meanings" (*"Bedeutungszusammenhang"*) or a "totality of references *(Verweisungsganzheit)*, which is constitutive for worldliness *(Weltlichkeit)* itself" (SZ, p. 76/107). The network of references to other words is a condition for the possibility of both the origination and the understanding of a work. At the same time, it forms the greatest obstacle to this understanding and constitutes its limitation. Thus, Heidegger writes the following:

> [T]he thinking of the modern epoch is much more difficult of access than the thinking of the Greeks, for the writings and works of the modern thinkers are structured differently, are more multilayered, are pervaded by tradition and are always engaged in the controversy with Christianity. (SG, p. 123)

On the basis of this "complicated state of affairs," philosophy runs the risk of becoming mere "idle talk" and being utterly unintelligible, which is to say, instead of *discovering (entdeckend)*, doing nothing but *covering over (verdeckend)*.

Because philosophy is a *construction (Konstruktion)* it can also be subjected to a *destruction (Destruktion)*, or better, to a de-construction, an activity through which the internal structure and the constitutive elements of factical philosophy are brought to light. Such activity is ultimately oriented toward trying to direct attention to the unthought (*das Ungedachte)* in thinking and to the unsaid in saying. The expressions "the unthought" and "the unsaid" are again somewhat ambiguous and it is not always easy accurately to distinguish the different meanings which these expressions can have for Heidegger. The unthought or the unsaid can be that which was never expressly thematized although it was continuously presupposed in (philosophical) thinking and which, indeed, *can* be thought and said. To this meaning belong the enduring, constantly returning, and historically changing structures of philosophy (such as its ontotheological character). The unthought or unsaid can also be that which is essentially, and always remains, unthought and unsaid, yet which at the same time is constitutive for every thinking and every saying. This unthought or unsaid never passes over into the realm of the thought or the said. It even waxes and becomes greater in the degree that more and more comes to be thought and said. Heidegger writes the following:

> The greater the work of thinking which claims a thinker..the richer will be the unthought in this work of thinking, that is, the richer will be that which comes forth first and only in this work of thinking as the not-yet-thought. This unthought, of course, has nothing to do with something which a thinker has overlooked or failed to master and which posterity, because it understands such matters much better, will have to make up for. (SG, pp. 123-24)

The unthought in a philosophy constitutes that philosophy and its greatness, just as the forgottenness of Being *(Seinsvergessenheit)* is constitutive for metaphysics and its greatness.

Attention can be directed toward the unthought in thinking only by way of an accurate analysis of what in fact happens when thinking takes place, when something comes to be thought. Attention can be directed toward the unsaid in philosophy only by way of an accurate analysis of what in fact takes place in and through philosophy, that is to say, by means of the *destruction*. Here it is a question of a *return into the ground of metaphysics* (WM, Foreword) or a seeking "after the condition for the possibility of the understanding of Being where this understanding is to be regarded as such" *(Grundprobleme, p. 399)*. It is a question of a *step back (Schritt zurück)* "out of metaphysics into the issuant and abiding essence of metaphysics" (ID, p. 47/51), "out of the already thought into an unthought whence what has been thought receives room to issue and abide in its essence" (ID, p. 44/48). "The *step back* does not refer to an isolated step of thinking, but rather to the whole style of the movement of thinking and to a long path" (ID, p. 46/50). It is a question of trying to get the factically existing philosophies and factically existing thinking before oneself as a *Gegenüber* and then to ask what is factically consummated there.

This is a central theme in the later Heidegger, but it is already fully and clearly expressed in *Grundprobleme der Phänomenologie*, for instance in those passages where Heidegger

speaks about Hegel. Heidegger says of Hegel that he saw everything which there is to see and thought everything which there is
to think. With Hegel, philosophy is "in a certain sense thought to
the end" (GP, p. 400/282). Heidegger remarks, moreover, that
Hegel was able to see so much "because he possessed an unusual
power over language and wrested the concealed things from their
hiding place" (GP, p. 226/159). According to Heidegger, however, Hegel did not pose the question about the light in which he
was able to see what he saw, and think what he thought; he did
not ask precisely what happens when something comes to be
spoken and what makes possible every understanding and every
understanding of Being. According to a striking and remarkable
formulation in *Die Grundprobleme der Phänomenologie*, to pose
such a question is "to question forth and beyond Being", it is "to
go forth beyond Being toward the light out of and into which Being itself comes into the luminous brilliance of an understanding"
(GP, p. 400/282-83). This formulation no longer appears in the
later Heidegger, but we gather from it that Heidegger perceives
himself to be dwelling in the neighborhood of Plato and his *epek-
eina tes ousias* (GP, pp. 401-405/283-86). This is, finally, the
problem of the *difference (Differenz)*, the *difference* which is and
remains essentially unthought and unsaid, but which at the same
time is constitutive for every thinking and saying, for every
understanding and every understanding of Being, and which
makes all this possible. This *difference* can only be 'discovered'
in an activity or a movement of thinking which, during the period
of *Being and Time*, Heidegger called "destruction."

(Translated by Wilson Brown)

NOTE

1. My quotations are from the following works of Heidegger:

Grundprobleme der Phänomenologie, Gesamtausgabe Bd.
24. Frankfurt a.M.: Klostermann, 1975. *(GP)*

Was ist das--die Philosophie? Pfullingen: Neske, 1956
(WiP)

Sein und Zeit. Tübingen: Niemeyer, 1957. *(SZ)*

Was ist Metaphysik? Frankfurt a.M.: Klostermann, 1955. *(WiM)*

Identität und Differenz. Pfullingen: Neske, 1957. *(ID)*

Prolegomena zur Geschichte des Zeitbewusstseins, Gesamtausgabe Bd. 20. Frankfurt a.M.: Klostermann, 1979. *(Prolegomena)*

Kant und das Problem der Metaphysik. Frankfurt a.M.: Klostermann, 1951. *(KM)*

Der Ursprung des Kunstwerks, in *Holzwege.* Frankfurt a.M.: Klostermann, 1972.

Holzwege, Gesamtausgabe Bd. 5. Frankfurt a.M.: Klostermann, 1977.

Was Heisst Denken? Tübingen: Niemeyer, 1954. *(WD)*

Erläuterungen zu Hölderlins Dichtung. Frankfurt a.M.: Klostermann, 1951. *(EHD)*

Der Satz vom Grund. Pfullingen: Neske, 1958. *(SG)*

The page numbers to the English translations have been added by the editor; in order to avoid the duplication of brackets a slash has been used to separate the page numbers of the original works and those of the translations.

2. Edmund Husserl, *Die Krisis der europäischen Wissenschaften und die transzendentale Phänomenologie: Eine Einleitung in die phänomenologische Philosophie* (The Hague: Nijhoff, 1962), p. 192; *The Crisis of European Sciences and Transcendental Phenomenology*, trans. David Carr (Evanston: Northwestern University Press, 1970), p. 188.

8. JOSEPH J. KOCKELMANS

BEING-TRUE AS THE BASIC DETERMINATION OF BEING

In this paper I wish to focus on some issues raised in Heidegger's interpretation of Aristotle's conception of truth. In so doing I hope to make a contribution to the problem of how Heidegger in 1930 conceived of the relationship between the question concerning the meaning of Being and the question of *a-letheia*.[1]

Heidegger's interest in the relationship between Being and truth (*aletheia*) was aroused by Brentano's book, *On the Several Senses of Being in Aristotle*, which Father Conrad Gröber, then pastor of Trinity Church in Konstanz, later bishop of Meissen in Saxony and still later archbishop of Freiburg, had given him in the summer of 1970.[2] In this book Brentano explained that according to Aristotle being is said in various ways; one of these is being in the sense of "being true."[3] The question of what *on hos alethes, ens tamquam verum*, being in the sense of true, precisely means and of how what-is is to be related to what-is-true occupied Heidegger thus from the very beginning.[4]

In 1925-1926 Heidegger gave a course in Marburg entitled *Logik: Die Frage nach der Wahrheit*,[5] in which he had planned to treat the question of truth both "historically" and "thematically." For reasons not known to me Heidegger changed the outline of the course and eliminated the part which was to present us with a systematic explanation of his own conception of the essence of truth,[6] in order to focus instead on the temporality of care and a retrieve of Kant's conception of time as found in the *Critique of Pure Reason*.[7] Yet the first part of the course deals with the relationship between logic and truth and contains a very detailed study of Aristotle's conception of truth and Heidegger's retrieve of its basic ideas.[8] We find here already a great number of insights which Heidegger later would employ in section 44 of *Being and Time*.

It is perhaps interesting to note in passing that in *Prolego-mena zur Geschichte des Zeitbegriffs*[9] which contains the lecture course of 1925 that preceded the course on logic, the question of truth is not discussed either, even though we find already in this "first draft of *Being and Time*" the most important sections of this work.[10] I have the impression that in 1925 Heidegger was not yet completely ready to communicate the results to which his "cri-tical" and "historical" studies concerning truth had led him.

Be this as it may, the first systematic treatment of the question of truth which we now have is found in *Being and Time*, section 44. Here Heidegger states that from time immemorial phi-losophy has circled around the connection between Being and truth. This is particularly evident in Parmenides, but we find the same conviction also in Aristotle who defined philosophy as *episteme he theorei to on hei on*, the science which contemplates being as being, and equally as *episteme tes aletheias*, the science of the truth.[11] Heidegger then continues that in view of the fact that a fundamental ontology is to lay the foundation for ontology which is to concern itself with Being, fundamental ontology must try to establish the relationship between truth and man, if we are to understand the reason why "Being necessarily goes together with truth and *vice versa*..."[12]

Heidegger's investigation itself is divided into three sections: 1) the traditional conception of truth and its foundation (truth as the correspondence between intellect and thing), 2) the primordial phenomenon of truth and the account of the derivative character of the traditional conception of truth (truth as *a-letheia*: Being-true as Being-uncovered and showing the truth as uncovering), and finally 3) the mode of Being of truth and its presuppositions (there is truth only insofar as *Dasein* is).[13]

If we compare the ideas developed in *Being and Time* with the result of the retrieve of Aristotle's conception of truth dis-cussed in the lecture course on truth of 1925, then it is clear that already in 1925 Heidegger in his account of the essence of truth no longer refers to an *intellectus archetypus*, although he maintains the distinction between logical and ontological truths. Furthermore, in speaking about truth, already in 1925 Heidegger considered the relation between the knowing and acting human being and the beings known in terms of a relation between revealing and being-revealed, discovering and being-discovered, and the term *a-letheia* is already interpreted accordingly. It is said already in 1925 that my comportment is true when I am able to show what the thing actually is; on the other hand, a being is said to be true to the degree that it is in non-concealment, revealed, discovered just as it is. To-be-true (*Wahrsein*) for a being means thus to be discovered as actually present-at-hand (*vorhanden, on energeiai*) in its constant and abiding coming-to-presence (*Anwesenheit*). However, it is important to note that neither in the 1925 lecture on truth nor in *Being and Time* is there any trace of the idea that Being itself, as this historically comes-to-pass as the open domain in which man and beings are related to one another, has an essential bearing on the question concerning the essence of truth.

The main differences between the 1925 reflection on truth and section 44 of *Being and Time* are 1) that the historical dimension of the issue which in *Being and Time* is merely hinted at is described in detail in the lecture course of 1925,[14] and 2) that Heidegger in 1925 develops his own conception of truth by means of a careful retrieve of the conception of Aristotle.[15] As for the latter, Heidegger first shows that Aristotle never defined truth in terms of the statement (*logos*), he then describes the transition from the hermeneutic *as* implied in primordial understanding to the apophantic *as* constitutive for the statement (*logos*); finally he turns to the problem of truth by mainly focussing on two questions: a) what are the conditions of the possibility of our statements being capable of being false (the hermeneutical situation of *Being and Time*), and b) what is the precise relationship between Being and truth (Theta 10).

In 1930 Heidegger again returned to the question of truth and its relationship to Being. First there is the essay *On the Essence of Truth*.[16] Then there is the lecture course *On the Essence of Human Freedom* which was delivered during the summer semester of 1930.[17] At present it is still not totally clear how all the problems of chronology are to be resolved. We know that the lecture course of 1930, in which there is a long section on the relationship between Being, truth, and coming-to-presence, was delivered between April and August, whereas the lecture on truth was delivered first in Bremen in October and then repeated in Marburg on December 5 and in Freiburg on December 11. Finally, the lecture was published in 1943. In 1943 Heidegger indicated that the final text is the text of the lecture held in Bremen, but that the original text was revised several times. There is no indication of the dates on which these revisions took place. Perhaps it is not unreasonable to assume that in preparing the 1930 version of the lecture on truth Heidegger made use of ideas discussed during the summer semester of the same year.[18]

Although it is my intention in this paper to focus mainly on a few issues explicitly raised in the course *On the Essence of Human Freedom*, I wish first to summarize briefly some basic ideas developed in the lecture on truth.

II.

It is now quite commonly accepted that the lecture *On the Essence of Truth* marks the transition from Heidegger's earlier to his later philosophy. In his effort to lay the foundation for metaphysics in a fundamental ontology Heidegger first tried to clarify the relation between the process of Being and the finitude of that being which understands Being, namely *Dasein*. In *Being and Time* he analyzed the mode of Being of *Dasein* phenomenologically and hermeneutically in order to find access to the problem concerning the meaning of Being. Subsequently he focussed more directly on Being itself and particularly on the problem of the truth of Being, in view of the fact that the meaning of Being is its truth. The growing importance of the problematic of truth, which can be discerned in the works that followed *Being and Time*, culminates in the essay on truth.[19]

The essay consists of ten sections. It can be divided into an introduction, in which Heidegger focusses on the tension between philosophy and common sense, two major parts (sections 1 through 3 and 4 through 7), a brief conclusion (section 8), and a note (section 9). In the introduction Heidegger discusses the question concerning the tension between philosophy and common sense, a tension which directly confronts us with the question of truth. In part one he takes his point of departure from the common conception of truth, namely truth in the sense of correspondence, and inquires into its presuppositions and its ground. In part two Heidegger explains why the essence of man is to be rethought in light of the outcome of the analysis contained in part one. It appears namely that the inquirer himself is changed because of the investigation. Thus part two provides us with an exposition of the essence of *Dasein* inasmuch as this must be understood in terms of the essence of truth. This exposition finally leads to a new conception of philosophy which concludes the essay. "The conclusion, however, does not constitute the end, in the sense of reaching the goal, but rather points forward to a beginning which remains to be made afresh."[20]

In the Introduction to the essay Heidegger writes that he will make an effort here to determine the very essence of truth, i.e., that which characterizes every kind of truth *as* truth, the ground of its inner possibility. According to ordinary common sense, however, such an investigation is superfluous; it makes no sense to inquire into the essence of truth, since nobody is really interested in the *nature* of truth. What interests us is the "*real*" truth which can give us a standard against the prevailing confusion of opinions and calculations. If a questioning concerning truth is necessary what we then demand is an answer to the question as to where we stand today. Yet, Heidegger observes, if we want to call for the "actual" truth, then we must realize that in calling for the actual "truth" we must already know what truth as such means.[21]

Section one begins with an analysis of the conventional conception of truth, according to which truth consists in the conformity between judgment and judged, *adaequatio intellectus et rei*. There is an ontological truth, in which the thing conforms to an intellect, and a logical truth, in which the intellect conforms to the thing. In the first case, the thing means every created thing, whereas the intellect referred to is the divine intellect. Thus this conception of the ontological truth "implies the Christian theological belief that, with respect to what is and whether it is, a thing, as created (*ens creatum*), *is* only insofar as it corresponds to the idea preconceived in the *intellectus divinus*, i.e., in the mind of God..."[22] In the second case, the issue is about the human intellect which must conform to the things which it knows. And it is claimed further that the proper place of truth in this case is the human intellect's act of judgment. It is important to realize that in this conception of truth the measure of truth in both cases lies in the correctness (*Richtigkeit*) of the conformity mentioned; untruth simply is non-conformity, or incorrectness.[23]

In Heidegger's view, this traditional conception of truth is undoubtedly correct and even important. Yet we must ask the question concerning the conditions which must be fulfilled in order to make this conformity possible, because it is in these conditions that the *essence* of truth is to be found.[24]

In developing his own conception of these conditions in the sections 2 through 8 Heidegger makes use of the historical investigations in which he had been engaged at least since 1925, as well of the thematic reflections as found in *Being and Time* and the lecture course *On the Essence of Human Freedom*. As for the latter, Heidegger was concerned there primarily with an effort to show that freedom, properly understood, indeed is the condition of the possibility of the manifestation of the Being of beings and thus also of *Dasein*'s comprehension of Being.[25]

Let us now turn to the lecture course on human freedom in an effort to resolve a few important problems which because of the ambiguity concerning the chronology of the lecture on truth alone cannot be resolved, namely, how did Heidegger in 1930 precisely conceive of the ontological difference and of the meaning of Being as such and how did he then conceive of the relationship between Being and truth?

III

On the Essence of Human Freedom is thus concerned with the relationship between truth and freedom. The second part of the course focuses on a retrieve of Kant's conception of transcendental and practical freedom,[26] whereas the first part tries to answer the question of how the problem of freedom precisely is to be related to the basic problem of philosophy, namely the question concerning the meaning of Being.[27] In the first part Heidegger explains, among many other issues, that the link between Being and freedom consists in truth. Heidegger substantiates this in part by showing how already in Greek philosophy the question of being as such (*on hei on*) led to reflections on *ousia, energeia,* and *aletheia.* Aristotle discussed these issues in *Metaphysics,* books E through *Theta.*[28]

In the reflections to follow it is important to keep in mind that in his analysis of the relevant themes of Aristotle's *Metaphysics,* Heidegger was completely familiar with the common interpretation of Aristotle's views by the leading Greek, medieval, and modern commentaries and that he was equally familiar with the scientific research concerning Aristotle's *Metaphysics* that has been done since the first part of the 19th century. Thus even though in 1930 Heidegger did not explicitly dwell on the "classical" and modern interpretations (with the exception of a few remarks on Schwegler, Jaeger, and Ross), the information provided in the lecture course of 1925 is here preunderstood.

Heidegger begins his reflection with a few brief remarks on *Metaphysics,* E, chapters 2 through 4. Aristotle states there (1026a33ff.) that the word "being" is used in many senses. In *Metaphysics* E, he mentions four ways in particular:

1) *to on kata ta schemata ton kategorion*: being according to the 'types' of the categories (whatness, quantity, quality, etc.);

2) *to on kata dunamin kai energeian*: being insofar as it is

potential or actual;

3) *to on kata sumbebekos*: being insofar as it happens to be such and so by accident;

4) *to on hos alethes*: being as that which is true. (1026b34ff.)

Heidegger then continues that according to Aristotle himself the last two senses of being are not to be discussed in metaphysics; only the first two senses are to be treated; the first in book *Z and H*, and the second in *Theta*, 1-9. The third sense is excluded because it does not mean Being as continuous presence; it seems that something that is by accident is something closely allied to non-being (1026b21). The fourth sense is excluded because it is just an affection of thought (*tes dianoias ti pathos*) (1028a7). Thus being in the sense of what is true has to do with thought, not with Being itself; it is to be discussed in logical treatises. Being in the full sense of the term is *ousia* taken as *eidos (Z)*. Later in *H* it is added that *he ousia kai to eidos energeia estin* so that being in the strictest sense is *on energeiai* (1050b2). But this means that *Theta* as that book of the *Metaphysics* where Aristotle speaks about being in the strictest sense of the term, defines Being finally in terms of *energeiai on*.[29]

But with this interpretation Theta 10 appears to be in direct contradiction. For there Aristotle claims: "...'being'...has senses according to the types of the categories and also according to potentiality and actuality...but being in the most proper sense (*to de kuriotata on*) is what is true..." (1051a35-b3). It is not only the case that Aristotle now explicitly begins to deal with being in the sense of what is true, but he even does so at the place where his *Metaphysics* reaches its summit, namely at the end of the books *E* through *Theta*. Furthermore, he *explicitly* states that being, taken in the most proper sense, is what is true.[30]

Because of this apparent contradiction, Heidegger says, many contemporary authors do not hesitate to claim that this chapter (*Theta*, 10) does not belong to the *Metaphysics* and is thus to be eliminated. It is a chapter taken from a logical text that by mistake was added to the last chapter of *Theta*. Thus, although these authors maintain that the ideas developed in this chapter are certainly authentically Aristotelian, they nonetheless also state that this chapter does not belong to *Theta*, because Aristotle has just proven there that *on* in the proper and strict sense is *energeiai on*; he can therefore now not claim that being in the sense of what-is-true is being in the most proper sense (Schwegler, Jaeger, Ross).[31]

To resolve this difficulty some authors reject *Theta* 10 altogether. Others have argued that *kuriotata* does not mean here "in the most proper sense," but rather has the meaning of "preferably" or even "usually." Heidegger does not deny that *kuriotata on* could have the meaning of "being in the usual sense of the term," but he nonetheless still feels that these authors are mistaken. *Theta 10* is an essential part of Aristotle's *Metaphysics* and it even constitutes philosophically its summit. The fact that Aristotle concludes the basic treatises of his *Metaphysics*, namely the books *E* through *Theta*, with a brief reflection on being in the sense of that-which-is-true has an important reason: it shows the decisive and fundamental significance of both Being and truth for the metaphysics of the ancient Greeks.[32]

Heidegger states again that according to Aristotle's explicit testimony in *Theta* 10 Being-true is the most proper Being of the beings and that in what-is-true as such the most proper essence of Being manifests itself. These claims are furthermore made in that book in which Aristotle constantly speaks about Being in the strict sense, namely *energeiai on*, namely book *Theta*.[33] *Theta* 10 furthermore delivers an explicit proof for the thesis that Being-true (*Wahrsein*) constitutes the most proper Being of the beings taken in the strict sense. Aristotle begins there with the question of how the being as being itself has to be a being in order to be capable of Being-true. In order to prevent misunderstanding, Aristotle himself adds that he is concerned here with *alethes on* in regard to the beings taken as things *(epi ton pragmaton)* (1051b2). Thus he wants to speak about the truth of beings, not about the truth of the *logos*, the statement or judgment. In other words, Aristotle himself states here explicitly that he is now no longer concerned with the logical truth of statements, about which in *E* 4 he had said that he wished to exclude it from metaphysics.[34]

Heidegger asks then how the leading philologists could misunderstand Aristotle's most basic claim. In his view this is due to the fact that they all assumed that one does not need to ask what Aristotle really meant by Being and by truth. They all assumed that it is evident what Being is and that truth obviously consists in the conformity of statement and state of affairs. What they thus did not realize is that in antiquity *a-letheia* meant non-concealment. Taken as such *a-letheia* is not characteristic of man's knowledge, but rather of the things themselves. Non-concealment is the non-concealment of beings. Statements are not true if truth is taken in the primary sense of non-concealment. Statements merely relate the manner in which humans let beings be true and preserve them as such *(aletheuein)*. A being does not *let* something be true *(aletheuei)*; it is *on alethes*. Thus *alethes* and *aletheia* are inherently ambiguous; they both refer to revealing and being-revealed. Knowledge reveals because beings are non-concealed.

Heidegger then turns to the opposite of truth: "Non-truth is not identical to hiddenness, but rather to dissimilation *(Verstellt-heit)*. A corresponding distinction is thus to be made also between falsehood and non-truth--beauty also belongs to non-truth--, but falsehood is found there where something is lacking as far as the truth is concerned, where one finds non-conceal-ment and yet mostly dissimilation, where something indeed gives itself and yet just passes itself off as something which in truth it is not."[35] I shall return to the implications of this shortly.

Be this as it may, the fact that Aristotle in *Theta* 10 speaks about the Being-true of beings is unquestionable in that he explicitly states this at the beginning of the chapter: "It is not because we are in-truth thinking that you are white, that you are white; it is because you are white, that we speak the truth when we say so" (1051b6ff.). *Man's aletheuein is thus founded on the alethes on*.

According to Heidegger we must thus say that Aristotle already made a distinction between logical and ontological truth. Yet in contradistinction to the Christian and modern tradition he does not interpret the ontological truth in terms of a relationship to an *intellectus archetypus*, i.e., to God's intellect; instead he defines ontological truth as non-concealment. Beings are true to the degree that they show themselves and manifest themselves.[36] And for this the human "intellect" suffices.

Aristotle formulates the basic problem in 1051b5-6: "But what then is, or is not, that which we say to be true or false?" When is that which is, something that is true? How must the Being of beings be in order that beings can be true in the sense of non-concealed and revealed? In Heidegger's view the answer can be given in the following theses: 1) When in regard to beings every possibility of non-truth is excluded in every possible respect. 2) This is so if Being-true belongs to Being. 3) The latter is so when Being-true constitutes what is most proper to Being as such. 4) Since Being is continuous coming-to-presence *(beständige Anwesenheit)*, the latter thesis is true if truth means nothing but the most proper coming-to-presence. The question of why *alethes on is kuriotata on* is an ontological, not a logical or an epistemological issue.[37]

On the pages 93-106 Heidegger explains his answer in minute details. I can here only mention the basic points not yet indicated in the provisional answer just given. Whereas the entire Greek and medieval tradition interprets *Theta* 10 to be dealing with the truth insofar as it is found in man's knowledge and then establishes that there is no truth in the simple apprehension of simple things, but that there is truth in the affirmative or negative judgment, there Heidegger claims that *Theta* 10 speaks about the Being of beings and that Aristotle here wonders how beings are to be in order that these beings can be true. Furthermore, Heidegger claims, the issue is not about simple apprehension and judgment, but about different modes of Being; Aristotle speaks about *sumbebekota* and about things that necessarily lie together *(sunkeisthai)*. Thus Heidegger translates 1051b9ff., as follows: Now, if whereas some things are always lying-together (this piece of chalk is material) and thus cannot be separated, and others are always separated and cannot lie together (the piece of chalk is spiritual), others again admit of both these possibilities (this piece of chalk is white), then to be means to lie-together and to be one, and not-Being is not-to-lie-together and to-be-multiple. To these two modes of being-one: *sunkeisthai* and *sumbebekenai* there belong two different ways of non-Being or Being-absent. To Being in the sense of *sunkeisthai* a determinate, possible non-Being belongs. The second mode of Being (that in the sense of *sumbebekenai*) is as such already in some sense a form of non-Being.

As for the latter it is clear that what can be true today could become not true tomorrow, in view of the fact that a being can be in this way today and in another way tomorrow. A piece of chalk can be white today and after it has fallen into an inkwell it can be blue or red. In this case truth becomes untruth, totally independently of man's judgment or opinion. The truth of the being by accident is unstable and not constant. And this is so in the final analysis because its Being is not a stable and constant coming-to-presence. The situation is completely different in the case of Being in the sense of *sunkeisthai*. Once a thing has been disclosed and uncovered in its what, it cannot be disclosed later in another manner; permanently and abidingly it is what it is. Yet the *sunkeimena*, too, are not free from *all* possible dissimulation *(Verstelltheit)*. This has been shown in section 7 of *Being and Time*. There are even different ways in which dissimulation can take place here. Thus Aristotle can say, "with respect to things which cannot be otherwise (the *sunkeimena*) it cannot happen that they are sometimes true and sometimes not true, but they are always either true or false" (1051b15f.). For these *sunkeimena* Being means continuous and constant coming-to-presence.[38]

But if all of this is so then it is also evident that what is not-composed (*adiaireta*, undivided, indivisible), not-put-together *(asuntheta)*, and simple is such that its Being is continuous and constant coming-to-presence. But then the simple in this sense is also that which is being in the most proper sense of the term. But this means that to be simple means to be true in the highest sense. The reason for this is that in this case Being-true can no longer be related to a possible not Being-true.[39]

Thus instead of interpreting Aristotle's reference to that which is simple as a reference to an absolute *akineton on*, or to the simple apprehension of the essence of a thing, as many commentators have done in the past, Heidegger interprets the reference to be one which is oriented to the Being of beings as such. But then it follows at once that for each being to-be-true *(Wahrsein)* means Being in the most proper and strict sense of the term. Heidegger explains his position in 1930 as follows.

The basic problem of philosophy is *ti to on*, what is being? The issue is here about that which constitutes a being as such, about that which constitutes its inner possibility, i.e., that from where and because of which it is capable of being what it is. The expression "from where" refers to *arche*, principle or ground, and *aitia*. Thus Aristotle can claim that "the more simple is more nearly a principle than the less simple *(mallon arche to haplousteron, Met.* K 1, 1059b35). The more we penetrate toward the simple, the closer we come to the principles. We may therefore say that the more original a form of knowledge is and the more original the revealment of what is revealed is, the more simple the principles will be *(Met.* E, 1, 1025b3ff.). Now the question concerning being as such *(on hei on)* belongs to our first knowledge; thus it is also the most simple in that it searches for what is the ground of being. But what is this? What is that which belongs to each being as such? This is *Being itself, auto to on, the being itself taken as such and merely with respect to its own Being*. Being itself is not that which once in a while belongs to beings, but rather that which belongs to each being continuously, above all, and as such. This is the reason why Being as such, simplicity, and unity can no longer be set apart from one another. Being in the sense of *ousia (Wesenheit)* is the most simple as such and this simple is the final ground of the possibility of every factual and conceivable being. It is this most simple which is also the most proper in each being.

That which in regard to each being is most proper, i.e., that which constitutes its continuous coming-to-presence, constitutes the ground of its possibility, is also that which is most true *(tas ton aei onton archas anagkaion einai alethestatas, Met. alpha* 1, 993b28f.). Thus Aristotle can then also say that *kuriotata on* is *alethes on.* This means that Being itself taken as *ousia* and as *energeiai on* is basically and mostly Being-true, Being-discovered and revealed. Heidegger can therefore conclude that, seen from the perspective of his more radical interpretation of *Theta* 10, it is now clear "that Being as such must be discovered and revealed a priori, continuously, and absolutely, if beings are to be discoverable as such and become open to determination."[40]

Heidegger explicitly admits here that this conclusion is the result of a destructive retrieve and that he is concerned here more with what Aristotle still left unsaid than with a passive repetition of what Aristotle did say already. Furthermore, it should be noted that Heidegger in this way defines the ontological truth without any reference to a divine intellect as *intellectus archetypus,* or to a simple apprehension as opposed to judgment. Finally, if the simple is merely the Being of beings taken as Beingness, then it is also evident that Being-true *(Wahrsein)* means the constant and continuous coming-to-presence of what actually is at hand *(die Anwesenheit des energeiai on).*[41]

Heidegger finally concludes his analysis by reiterating that 1) the problem about truth discussed in *Theta* 10 has nothing to do with the issue of the so-called logical truth; Aristotle is here concerned neither with logical nor with epistemological issues. 2) The question concerning the Being-true of beings is unfolded here as the fundamental question concerning the most proper mode of the Being of beings. 3) Thus instead of just being an appendix that someone "by mistake" added to book *Theta,* chapter 10 constitutes the genuine summit of Aristotle's *Metaphysics* as a whole, and of book *Theta* in particular.[42] 4) What Aristotle here says about Being-true is in complete harmony with the basic conception of truth as we find it in the Greek world of that time. Being-true was for the ancient Greeks the continuous, constant, and pure coming-to-presence.[43]

IV

We must now finally turn to the question of where these reflections have led us. It seems to me that if one compares the 1925 reflections on truth with those of *Being and Time* and those of 1930, it is quite clear that between 1925 and 1930 there is very little development in Heidegger's thinking with respect to his conception of truth. Already in the 1925 lecture course Heidegger had made it clear that he wanted to show that for Aristotle Being means *ousia, Anwesenheit, energeia (Vorhandensein,* Being actually present-at-hand), *alethes (Wahrsein,* Being-true). It is stated there also that *Theta* 10 is an essential part of Aristotle's metaphysics, and that it even constitutes its high point. In addition to a logical conception of truth which Aristotle usually discusses in his logical treatises *(de Interpretatione),* there is also an ontological conception of truth. This is unfolded mainly in *Metaphysica, Theta* 10. Yet this chapter should not be interpreted in such a way that the interpretation implies a reference to an *intellectus archetypus;* furthermore, the common interpretation according to which Aristotle is concerned here with the distinction between the manner in which the essences of things are immediately present to us in simple apprehension and the manner in which we unfold our knowledge of these essences in statements is inadequate. Already in 1925 Heidegger made the claim that for Aristotle Being, taken in the proper and strict sense, for each being means Being-true. This obviously also implies some relation to an intellect; yet the "intellect" implied here is man's understanding. This reference to *Dasein*'s understanding explains why Heidegger could write: only as long as *Dasein* is is there truth.

For we should not forget that coming-to-presence *(Anwesenheit)* essentially implies a *präsentieren* or *gegenwärtigen,* some form of making-present, and a Being-present-for. The ontological truth still implies some form of *Verhalten,* comportment on the part of *Dasein.* Thus to be as being-revealing on the part of *Dasein* is and remains still a condition for the revealment of what manifests itself. Due to the "hermeneutic situation" this revealment is always affected by negativity so that untruth belongs to the essence of truth. Yet, in the final analysis untruth is inherent in the Being of beings. Finally the fact that the ontological truth implies an essential reference to *Dasein*'s comportment as well as the fact that Being taken in the strict sense means the Being-revealed of what constantly and continuously comes-to-presence explain the essential link between Being and time, a link which is explained so convincingly in *Being and Time* but is mentioned in both 1925 and 1930.

At the present time I have no evidence for the thesis that in 1930 Heidegger had already surpassed this position in an essential respect. Nowhere is there any explicit reference to what later will be called Being itself as the open domain in which the ontological difference between a being and its Being can come-to-pass. The difference stressed here is thus not the transcendental ontological difference between Being itself *(einai, der Sinn von Sein)* and the Beingness of a being *(ousia, die Seiendheit eines Seienden)*, but rather the categorial ontological difference between a being *(on)* and its Being *(ousia, Seiendheit)*.[44] If there is a genuine development in Heidegger's thinking between 1927 and 1930 it consists in the fact that in 1927 it is said that truth in the primary sense consists in uncovering, whereas truth in a secondary sense consists in Being-uncovered. When *Dasein* concerns itself with beings these beings "become that which has been uncovered. They are 'true' in a second sense. What is primarily 'true'--that is, uncovering--is *Dasein*. 'Truth' in the second sense does not mean Being-uncovering..., but Being-uncovered..."[45] In 1930 this thesis has been reversed insofar as Heidegger there claims that Dasein's comportment reveals because and insofar as the beings are non-concealed.[46] Presumably the reason for this change was that the position of *Being and Time* clearly remains within the framework of modern metaphysics according to which the Being of beings is posited by man (will to power). Yet even in this context Being itself is not yet mentioned.

I therefore assume that what is usually called *die Kehre* was not yet made explicit in Heidegger's thinking in 1930. The reversal of thought must thus have been made explicit some time between 1930 and 1943. Yet because of the reason given in 1930 Heidegger certainly was already on the way to the turn.

NOTES

1. Cf. Jean Beaufret, "Heidegger et le problème de la vérité," in *Fontaine*, 63(1947), 758-785; Walter Biemel, *Martin Heidegger in Selbstzeugnissen und Bilddokumenten* (Reinbek: Rohwolt, 1973); William J. Richardson, *Heidegger. Through Phenomenology to Thought* (The Hague: Nijhoff, 1963), pp. 211-254.

2. Thomas Sheehan, "Heidegger's Early Years: Fragment for a Philosophical Biography," in *Heidegger: The Man and the*

Thinker, ed. Th. Sheehan (Chicago: Precedent Publishing Inc., 1981), 3-19, p. 4.

3. Franz Brentano, *On the Several Senses of Being in Aristotle,* trans. Rolf George (Berkeley: University of California Press, 1975), pp. 3-5, 15-26.

4. Martin Heidegger, "Neuere Forschungen über Logik," in *Literarische Rundschau für das katholische Deutschland,* 38(1912) 465-472, 517-524, 565-570, p. 269.

5. Ed. Walter Biemel (Frankfurt: Klostermann, 1976).

6. *Ibid.,* p. 26; cf. p. 417.

7. *Ibid.,* pp. 197-269, 269-415.

8. *Ibid.,* pp. 19-25, 127-161, 162-195.

9. Martin Heidegger, *Prolegomena zur Geschichte des Zeitbegriffs,* ed. Petra Jaeger (Frankfurt: Klostermann, 1979).

10. Including sections 42, 43, 45, and 46; note that section 44 is missing here. Cf. Martin Heidegger, *Sein und Zeit* (Tübingen: Niemeyer, 1963); *Prolegomena,* pp. 293-306, 417-420, 421-423, 424-431.

11. Aristotle, *Metaphysics, Gamma,* 1 1003a21 and *alpha,* 1, 993b20.

12. SZ 213.

13. SZ 214-230.

14. *Prolegomena,* 99-125.

15. *Ibid.,* pp. 127-195.

16. Martin Heidegger, *Vom Wesen der Wahrheit* (Frankfurt: Klostermann, 1961).

17. Martin Heidegger, *Vom Wesen der menschlichen Freiheit. Einleitung in die Philosophie,* ed. Hartmut Tietjen (Frankfurt: Klostermann, 1982).

18. WW, p. 4; Walter Biemel, *op. cit.,* p. 152.

19. William J. Richardson, *op. cit.,* pp. 26-27.

20. Walter Biemel, *op. cit.,* pp. 26-27.

21. WW, 5-6.

22. WW, 30.

23. WW, 8-9.

24. Richardson, *op. cit.,* p. 213; Biemel, *op. cit.,* pp. 77-79.

25. WmF, "Nachwort des Herausgebers," p. 307.

26. WmF, 139-297.

27. WmF, 17-138.

28. WmF, 39-109.

29. WmF, 77-79.

30. WmF, 80.

31. WmF, 80-84.
32. WmF, 84-87.
33. WmF, 87.
34. WmF, 87-88.
35. WmF, 91.
36. WmF, 91-92.
37. WmF, 92.
38. WmF, 98-99.
39. WmF, 99-100.
40. WmF, 103-104, cf. 102-104.
41. WmF, 104-108.
42. WmF, 107.
43. WmF, 108.
44. Joseph J. Kockelmans, "Destructive Retrieve and Hermeneutic Phenomenology in 'Being and Time'," in *Research in Phenomenology*, 7(1977), 106-137, pp. 115-116.
45. SZ 220.
46. WmF, 90-91.

9. WILLIAM J. RICHARDSON

HEIDEGGER AND THE QUEST OF FREEDOM*

What characterizes the age of Christian renewal is the quest
of freedom. For man in our time has a deeper awareness than
ever before of the mystery of his own liberty. The reason may lie
in the historical moment itself; for it seems that at this stage of
his development man is being invited to assume more and more
responsibility for the direction of the evolutionary process out of
which he himself has emerged. In any case, the Church herself
feels the same stirrings in her own members; for in any given
area of crisis in the postconciliar age--and most dramatically perhaps in the area of morality--the quest of freedom plays a significant, sometimes decisive, role. It is the purpose of these pages
to raise the question as to whether the thought of Martin Heidegger can offer any light to that quest, no matter how trammeled
with darkness that light may be.

To be sure, the question of freedom is not the specifically
Heideggerean question. Still less is he concerned with the question of morality (and least of all a "new" one). Rather, as we all
know, his question is the question of Being. But the Being-question itself brings Heidegger to grips with the notion of freedom
time and again along the way, so that it is not a distortion for us
to examine his thought under this aspect. And once we come to
grips with the problem of freedom, surely the question of morality
is not far away. Let us follow this general sequence of thought as
we proceed.

The basic orientation of Heidegger's effort at posing the
Being-question is by now fairly common knowledge. How he came

*From William J. Richardson, "Heidegger and the Quest of
Freedom," *Theological Studies,* 28 (1967), pp. 286-307. Reprinted
with permission of the author and the Editor of *Theological
Studies.*

to the question he has made clear himself. At the age of eigh-
teen, when he was at the academic level of about a college sopho-
more, a priest-friend gave him a copy of Franz Brentano's doctoral dissertation *On the Manifold Sense of Being in Aristotle*,
where "Being" translates the German *Seiendes* and the Greek *on*,
both signifying "that which is." He describes the experience in a
familiar passage:

> ...On the title page of his work, Brentano quotes Aristotle's
> phrase: *to on legetai pollachos*. I translate: "A being becomes
> manifest (i.e., with regard to its Being) in many ways."
> Latent in this phrase is the *question* that determined the way
> of my thought: what is the pervasive, simple, unified deter-
> mination of Being that permeates all of its multiple meanings?
> ...How can they be brought into comprehensible accord?
>   This accord cannot be grasped without first raising and
> settling the question: whence does Being as such (not merely
> beings as beings) receive its determination?[1]

The Being-question, then, was posed early. Heidegger goes
on to list some of the forces that influenced him as he began to
elaborate it. The first he mentions is Edmund Husserl:

> Dialogues with Husserl provided the immediate experience
> of the phenomenological method that prepared the concept of
> phenomenology explained in the Introduction to *Being and
> Time* (#7). In this evolution a normative role was played by
> reference back to fundamental words of Greek thought which
> I interpreted accordingly: *logos* (to make manifest) and
> *phainesthai* (to show oneself).[2]

Husserl, then, supplied him with a method. But what he
does not mention, yet what seems equally decisive for the young
Heidegger, was the Husserlian experience that for a phenomenologist a "being" is that which appears, is present as meaningful to
him. It would follow that the Being of such a being would be the
process that *lets* such a being appear to the philosopher and be
present as meaningful for him.
    Another early influence, no doubt under the aegis of
Brentano, was Aristotle--but in a rather unusual way: "A re-
newed study of the Aristotelian treatises (especially Book IX of
the *Metaphysics* and Book VI of the *Nicomachean Ethics*) resulted
in the insight into *aletheuein* as a process of revealment, and in

the characterization of truth as non-concealment, to which all self-manifestation of beings pertains...."[3] In other words, there is evident even in these early years a correlation between Being, conceived as a process of revelation by which beings appear, and truth, conceived as a process of non-concealment. For by Being a being becomes revealed, i.e., the veil (*velum*) of obscurity that conceals it is torn aside (*re-*). In Greek, the word for conceal-ment is *lethe*, and privation is signified by an alpha prefix. When a being becomes re-vealed, it becomes un-concealed (*a-lethes*), i.e., (for the Greeks) "true." Being, then, is conceived as a process by which non-concealment (*a-lethei-a*: truth) comes about. By the same token the being in question may be conceived as "liberated" from concealment and Being (*aletheia*) a process of liberation, of making beings free. From the beginning of Heideg-ger's way, then, Being, truth (*aletheia*), and freedom are insepa-rably intertwined.

Once this basic insight is clear, it is easy to understand that the treatment of the problem of freedom will run parallel--at least by implication--to the problem of Being, and follow the same vagaries along the way. For the sake of clarity, then, let us examine the notion of freedom first in the early Heidegger (let us call him Heidegger I), then in his later period (Heidegger II), and conclude with some questions of our own.

I

By Heidegger I, we understand the Heidegger of *Being and Time* and of those earlier works, prior to 1930, which share the same perspectives. Now there is, to be sure, a discernible con-ception of freedom in *Being and Time* (1927), but amid the welter of analyses there it remains in the oblique. Perhaps we can get to the heart of the problem more incisively if we begin with Heideg-ger's thematization of the problem of freedom in the much shorter (though hardly more readable) essay *On the Essence of Ground* (1929). There we find as explicit a statement as this: "...Tran-scendence to the World is freedom itself...."[4] For Heidegger I, then, transcendence and freedom are somehow one.

Heidegger is perfectly aware, of course, that his remark is startling, and he passes immediately to the defensive. The tradi-tion conceives of freedom as one form or another of "spontanei-ty," i.e., as a type of causality by which the self initiates [something] of and by itself *(Von-selbst-anfangen)*. This, how-ever, is a purely negative conception of freedom, he claims, in the sense that the self is conceived as a cause whose causality is not determined by some other cause. To explain such a conception positively, one would have to explain ontologically (1) the nature of the self, and (2) the fundamental process character *(Geschehenscharakter)* of its structure, in order to explain how the self *can* initiate anything at all. Now "...the selfhood of the self that already lies at the basis of all spontaneity consists in transcendence...."[5] What, then, is the nature of the self conceived as transcendence? In what does its process-character consist? By what right can this be identified with freedom?

"Transcendence" is not a specifically Heideggerean word. Aside from *On the Essence of Ground*, we find it thematized in his own name only in the closing section of *Kant and the Problem of Metaphysics*, where Heidegger's purpose is to make clear to the reader the close relationship between his own problematic (already developed in *Being and Time*) and that of Kant, at least as he understands Kant.[6] As he reads Kant, the purpose of the *Critique of Pure Reason* was not to construct a theory of knowledge but to lay the foundation for metaphysics (i.e., the *metaphysica specialis* of the Leibniz-Wolff tradition). Insisting on the finite character of human knowing, according to which the knower does not create the objects of his knowledge but must receive them, Kant probed the a priori (i.e., pre-experiential) conditions of the possibility of this knowing. Now if, for the finite knower, the givenness of being-to-be-known is itself conceived a priori, then there must be built into the structure of the knower himself a pre-experiential comprehension of their structure as beings, i.e., of their Being, which may be conceived as a sort of domain or horizon within which these beings *can* be encountered and known. This a priori horizon of encounter is what Heidegger in Kant's name calls "transcendence."[7] Heidegger's own explanation can hardly be improved upon:

> A finite knowing essence can enter into comportment with a being other than itself which it has not created, only when this already existing being is in itself such that it can come to the encounter. However, in order that such a being as it is can come to an encounter [with a knower], it must be "known" already by an antecedent knowledge simply as a being, i.e., with regard to its Being-structure....A finite [knower] needs [a] fundamental power of orientation which permits this being to stand over in opposition to it. In this original orientation, the finite [knower] extends before

himself an open domain within which something can "corre-spond" to him. To dwell from the beginning in such a domain, to institute it in its origin, is nothing else than the tran-scendence which characterizes all finite comportment with beings....[8]

How Heidegger justifies his interpretation of Kant's endeavor need not concern us here. At the moment it is important only to see how the word "transcendence," thus understood, is trans-posed into his own problematic. "...Man is a being who is immersed among other beings in such a way that the being that he is not as well as the being that he is himself have already become constantly manifest to him...." So far, this is nothing but what in Kant he calls "transcendence." But he adds immediately: "...This manner of Being [proper to] man we call existence...."[9] For Heidegger I, if transcendence and freedom are one, so too are transcendence and existence.

In *Being and Time* "existence" is described as the Being of *Dasein*.[10] *Dasein*, of course, is the name chosen by Heidegger to designate the nature of man insofar as he is characterized before all else as endowed with a special comprehension of Being that permits him to discover and name beings as what they *are*. Exis-tence, thus understood, is later on written as ek-sistence, to suggest more clearly its fundamental nature. In other words, by reason of its Being *Dasein* stands *(-sistit)* outside of *(ek-)* itself and toward Being, the lighting-process by which beings are revealed. We may add, too, that in the phenomenological analysis of *Being and Time* Being reveals itself as the horizon of the World, so that *Dasein*'s openness towards Being can be described as to-be-in-the-World. In any case, it becomes perfectly clear that whatever the justification of its Kantian antecedents, tran-scendence for Heidegger means the same thing as existence, *Dasein*, and to-be-in-the-World: it designates *Dasein*'s structural comprehension of Being by reason of which *Dasein* can pass *(-scendit)* beyond *(trans)* all beings, including itself, to the Being of beings by which they are revealed to it. It is this passage that characterizes *Dasein* as a self and accounts for the fact that its fundamental structure is not that of a substance but of a process *(Geschehen)*. So far, so good. But by what right is such a process called freedom?

Before we can understand this clearly, we must review the essential elements of the phenomenological analysis of *Dasein* as it develops through *Being and Time*. In the briefest possible terms, we may say that the phenomenological analysis reveals *Dasein* to be transcendence that is finite, whose ultimate meaning is time.

Dasein is transcendence. This appears from the close analysis of what it means to-be-in-the-World. First Heidegger examines the World and discovers it to be not simply a horizon within which beings are encountered but a matrix of relationships within which they have meaning. Then he examines what it means to-be-*in* such a World. Fundamentally it means to disclose the World, and by reason of this disclosure beings within the World are disclosed to *Dasein*. Heidegger finds three components of this disclosure of the World through *Dasein*'s in-being. The first he calls "com-prehension," not in any intellectual sense but as a seizure (-*prehendere*) by *Dasein* in and as itself (*cum-*) of the pattern of meaningfulness that the World supplies. The second he calls "the ontological disposition" *(Befindlichkeit)*, that component of *Dasein*'s structure by which it is affectively disposed to other beings, responds to them, reverberates with them in all its various moods. Finally, the third component of *Dasein*'s In-being in the World Heidegger calls "logos" *(Rede)*. By this he understands that element in *Dasein* by reason of which *Dasein* can articulate its presence in and to the World through language. This complex structure by which *Dasein* is in-the-World is what the phenomenological analysis discovers in transcendence. We should add here perhaps that Heidegger insists that *Dasein* is never solitary in the World. It ek-sists with other *Daseins* (*Dasein* is *Mitdasein*), and this interlacing structure is the basis of all empathy.

Be that as it may, transcendence is finite, i.e., it is limited by many different kinds of "not." To begin with, *Dasein* is not its own master--it does not create itself but finds itself as a matter of fact in the World. Heidegger calls this *Dasein*'s "thrownness." Furthermore, *Dasein* is not independent of other beings but is related to them and in this reference depends on them to be what it is. Again, this referential dependence goes so deep that *Dasein* tends to become absorbed in other beings, becomes fallen among them ("fallenness") to such an extent that it tends to be oblivious of its openness to Being, to forget its true self. In its everyday condition, *Dasein* is normally victim of this fallenness, caught up in the throes of what everybody else says and does. Heidegger discerns this condition graphically as a subservience to "everybody else" *(das Man)*.

Another kind of "not" that marks the finitude of *Dasein*'s transcendence is the fact that Being itself, when considered in terms of beings, can only be experienced as not-a-being, Non-being *(Nichts)*. But the deepest "not" of all is the fact that *Dasein* cannot be forever, it is destined to die. So deep is this negativity of death that its sign is upon *Dasein* from the beginning--not as an event still to come but as already circumscribing the finite *Dasein*. As soon as it begins to *be*, it begins to be *finite*, and the supreme finitude that circumscribes it from the beginning is death. From the first moment of ek-sistence, then, *Dasein* is Being-into-death. The sum total of all these different types of finitude Heidegger calls "guilt." Because it is finite and inasmuch as it is finite, *Dasein* is ineluctably guilty.

Such, then, are the ingredients of the self as finite transcendence. Thrown among beings, it is open to their Being, yet trammeled with finitude, i.e., guilt. But how are these elements experienced in their unity, as pertaining to a single self? It is here that Heidegger describes the phenomenon of anxiety as revealing the true nature of the self. Anxiety is a special mode of the ontological disposition, an affective, nonrational attunement within us. It is different from fear, because fear is always an apprehensive response to something--like a dentist's drill--a being. But in anxiety the self is not anxious about any one thing but about no-thing in particular, about nothing! Yet not absolutely nothing, rather about "something" quite "real" that is still not a thing like other things, nor is it situated here nor there nor anywhere. Anxiety reveals *Dasein* as exposed to "something" that is no-thing and no-where. At this moment the things that have a "where" around us seem to slip out of our grasp, lose their meaningfulness. We are no longer at home among them. We are alienated from them, as we say--we are alienated, too, from "everybody else," from *das Man*, with all that they do and say. We discover that there is another dimension in life than the everyday one, a new horizon of which we are ordinarily unaware, yet within which and toward which we truly ek-sist, whether we call this horizon simply the No-thing *(Nichts)*, the World, or even Being itself. Through the phenomenon of anxiety, then, the self becomes aware of itself as a unified whole--related to beings within the World, yet open to Being, the World as such--aware, too, of the possibility of accepting the fact that this is what it is (finite transcendence), or of running away from the truth, refusing to know anything except what "everybody else" knows. In other worlds, the phenomenon of anxiety reveals to *Dasein* the possibility of choosing to be authentic or not.

But anxiety as such goes no further. It reveals *Dasein* to

itself but as such it does not call upon *Dasein* to make the choice to be true to itself. Yet there is such a voice that calls to *Dasein* out of its very depths--a voice that invites *Dasein* to be liberated from the thraldom of "everybody" and accept itself as finite transcendence, as openness to Being, shot through, as it is, with ontological guilt. This, for Heidegger, is the voice of conscience. To heed this voice means to say "yes": yes to its own transcendence--that is, to the fact that it will always be alienated from "everybody" to the extent that its true abode is not simply the level of beings alone but the domain of Being itself; yes to its own finitude, not as if this meant blind surrender to a tragic fate but simply a tranquil resignation to the fact that it is no more than it is. *Dasein* says "yes" to itself by what Heidegger calls the act of "resolve" *(Entschlossenheit)*, the moment when it achieves authenticity.

Dasein is finite transcendence, whose ultimate meaning--i.e., the ultimate source of its unity--is time. As transcending ek-sistence, *Dasein* is always coming to Being, i.e., Being is coming to it. This coming is *Dasein*'s future. But Being comes to a *Dasein* that already is. This condition of already-having-been is *Dasein*'s past. Furthermore, Being as it comes to *Dasein* renders all beings present as meaningful to *Dasein*. This presence is *Dasein*'s present. Future-past-present, these are the components of time. What gives unity to *Dasein*, then is the unity of time. To achieve authenticity precisely as temporal, *Dasein* must accept itself as essentially temporal--yes, and as historical, too.

There is much more to say, of course, but we must stop here if we are going to say anything about the question of freedom. In what sense does Heidegger maintain that to be truly authentic is to be truly free? In the sense that to be one or the other is to be true. What, then, does he mean here by truth?

We say that a statement is true when it expresses a judgment that is conformed to a situation of fact--in other words, when the judgment so judges a situation to be as it *de facto* is. But what guarantees this "so...as" relationship? Is it not the discovery by *Dasein* that the situation is as it is judged to be? More fundamental than conformity is this process of discovery of beings as they are, in their Being. But this process in *Dasein* which discovers the Being of beings--what is it but the comprehension of Being in *Dasein*--in other words *Dasein*'s ek-sistence, transcendence itself?

This process of discovering, which is *Dasein*'s transcendence, is the origin of truth as conformity, i.e., original truth.

That is why Heidegger can say that *Dasein* is "in the truth." But *Dasein*'s transcendence is finite, it is permeated by a multiple "not." For that reason the coming to pass of truth--truth in its origin, original truth--is likewise pervaded by a "not." Consider, for example, that aspect of *Dasein*'s negativity which we called "fallenness," i.e., *Dasein*'s built-in drag towards beings that propels it towards inauthenticity by inclining it to become a slave of "everybody" (*das Man*) and forget its privilege of transcendence. The process of original truth, too, is fallen among beings. This means that the discovery of beings is always somehow askew. They are discovered, to be sure, but always inadequately and drop back immediately into their previous hiddenness. For *Dasein* to apprehend a being *(ergreifen)* is simultaneously to misapprehend it *(vergreifen)*; to uncover *(entdecken)* is to cover up *(verdecken)*; to discover *(erschliessen)* is to cover over *(verschliessen)*. This condition of undulant, inescapable obscurity Heidegger calls "untruth." "...The full...sense of the expression '*Dasein* is in the truth' says simultaneously '*Dasein* is in the untruth'...."[11] And why? Because transcendence is finite.

Clearly, then, the coming to pass of finite transcendence is the coming to pass of truth in its origin. Now if *Dasein* achieves authenticity through that gesture of self-acceptance that is called "resolve," then resolve must be also the eminent mode of truth-- but also of untruth. In other words, if by resolve *Dasein* accepts the finitude of transcendence, it simultaneously consents to the finitude of truth. "...[*Dasein*] is simultaneously in truth and untruth. This applies in the most 'authentic' sense to resolve as authentic truth. [Resolve] authentically makes untruth its very own...,"[12] i.e., accepts the inescapable finitude of the transcendence which is the basis of truth.

But to do this is to become free. How? In *Sein und Zeit*, Heidegger uses two formulae with regard to the achieving of freedom. He speaks of "laying free" and of "becoming free." What he means by "laying free" becomes clear when we recall what he means by phenomenology. As we saw, it means *legein* (to let-be-seen) *ta phainomena* (beings whose nature it is to appear). But why should we have to make a special effort to let-be-seen these beings, unless these beings, in appearing as what they are, somehow conceal themselves as what they are? The effort to let them be seen, then, is an effort to liberate them from the obscurity that enshrouds them as what they are--to let them be free in truth. In truth! Recall what we know of the finitude of original truth, namely, that *Dasein* is in the untruth. As a result, the beings that *Dasein* illumines by reason of its comprehension of their Being-structure are so contaminated with negativity of this illumination that they conceal themselves as they reveal themselves. To let them be seen as what they are means to liberate them as far as possible from this concealment, in order that they may be manifest as what they are in truth. Truth must be wrested *(abgerungen)* from them; they must be torn away *(entrissen)*, robbed *(Raub)* from concealment in order that they may be manifest as what they are in truth. This is the sense Heidegger gives to the alpha prefix in *aletheia* here. It suggests the privation of, or liberation from, concealment. To lay something free, then, means to liberate it from obscurity—to let its truth come-to-pass.

What, then, does it mean to become/be free? The terminology Heidegger reserves to *Dasein* itself. As a matter of fact, the expression is used in two ways, and we might see in them two successive moments of the process by which *Dasein* lays its self free. The first moment of freedom occurs when *Dasein* is startled out of the complacency of its everyday absorption in beings and realizes for the first time that by its comprehension of Being it passes beyond these beings (including itself) to the process that lets them be (manifest). This occurs in the moment of anxiety when all beings seem to slip away from *Dasein* and leave it exposed to the "something" that is no-thing, the horizon of the World. In this moment *Dasein* has been laid free, liberated from the obscurity that had hitherto held captive the structures of its own transcendence. In this moment *Dasein*'s existence is wrested from (alpha prefix) the concealment *(lethe)* that held it prisoner; it is then clearly a moment of truth *(aletheia)*.

But only the first moment of truth, for it is only the first moment of freedom. "Anxiety," says Heidegger, "reveals in *Dasein*...[its] being-free-for [*Freisein für*] the freedom of choosing its self [*die Freiheit des Sich-selbst-wählens*]..."[13] In other words, this first moment of freedom makes possible a second moment in which it can choose to accept its self as transcendence that is finite, or to refuse its self by trying to run away from the awesome privilege of transcendence in yielding to the seduction of being one with "everybody else." In other words, it is free to choose between authenticity and inauthenticity. If it chooses to be inauthentic, it becomes a slave to the world of "everybody." If it chooses to be authentic, then, and only then, does it become authentically free. This happens, as we saw, when *Dasein* heeds the voice of conscience, calling it to achieve its self. "...In comprehending this voice," says Heidegger, "*Dasein* is attentive to the most characteristic potentiality of its existence. It has [thereby] chosen its self."[14] This choice is its resolve. In it *Dasein* liberates its self unto its self, achieves its self in authenticity, becomes authentically free.

For the early Heidegger, then, freedom is conceived fundamentally as achievement--achievement of the self. In all this the essential is to see that the primary sense of freedom is liberation in the sense of *aletheia*, the coming-to-pass of truth; that this comes-to-pass through the structure of *Dasein* as transcendence, ek-sistence, openness to Being-as-such; that *Dasein* itself brings the process to fulfillment when it achieves authenticity through the gesture of resolve.

Do we have the right to transpose any of this into terms of morality? As far as Heidegger is concerned, absolutely not. He conceives his question about Being (and about man only insofar as man has a built-in comprehension of Being) as far more radical than any question about the "oughtness" of human acts. We catch the spirit of his enterprise when we recall his insistence upon how Kant's three classic questions (1. What can I know? 2. What ought I to do? 3. What can I hope for?) are ultimately reduced to the fourth, which is the most fundamental of all: What is man [and, indeed, in his finitude]?[15] In raising a question about the Being of finite *Dasein*, then, Heidegger feels that he is getting deeper than the ethical problem as such. This viewpoint comes sharply into focus when he is dealing with the question of *Dasein*'s guilt. Though this notion normally appears in the context of morality, for Heidegger it expresses *Dasein*'s ontological "indebtedness," i.e., the sum-total of its finitude, and nothing more. But as such, it remains an ontological condition of possibility for moral action:

...This essential condition of being guilty is in an equally original way the existential condition of possibility for "moral" good and evil, i.e., for morality as such, and its possible matter-of-fact derivations. Morality cannot be what determines the original condition of guilt, because [morality] already of itself supposes [guilt].[16]

At best, then, Heidegger himself is dealing here only with the ontological structures that will be operative in any moral life, and these only insofar as they are part of the process of transcendence. But once this is said, is it possible for someone else who starts with a different experience--whether philosophical or religious--to legitimately utilize these Heideggerean structures to articulate his own experience, without claiming that the result is Heideggerean in any way other than that of inspiration?

If so, then all that is implied in the concept of authenticity might be very helpful. Fundamentally this means a free acquiescence to the finitude of truth which comes-to-pass through transcendence. Does this suggest a possible new way of speaking about conformity to moral law, or more specifically to so-called "natural" law, that would be correlative with the achievement of human liberty rather than a restriction of it? If by "natural" law we understand, grossly speaking, the law for man's action inscribed in his "nature," the "nature" of man in Heideggerean terms *(Wesen)* is obviously existence, transcendence, i.e., the finite process of original truth. As transcendence, *Dasein* is project of the World and therefore of its own potentialities as to-be-in-the-World. But the potentialities are constricted because transcendence is thrown into the matter-of-fact situation in which it finds itself. Thus, "thrown," *Dasein* is given over to itself to be. Truth *(aletheia)*, therefore, though illuminated through *Dasein*, is nonetheless *given* to *Dasein* to accomplish through its gesture of free acceptance. May we find here the ingredients of law-as-norm, whereby the law to be accomplished is essentially the process of *aletheia* and therefore precisely *as* law also liberation?[17]

Again, may we find some way of speaking about law-as-command, whereby the imperative character of the moral ought finds its foundation in the ecstatic nature of ek-sistence itself as drive-toward-Being? In this sense conscience, as the existential component called "logos," would let-be-seen *by* the self the finite process of *aletheia as* the self, and by this very fact call from *Dasein* on its ontological level to *Dasein* on the ontic level, lost in the distractions of *das Man*, and summon it to be true to its self--both ontic and ontological at once. Such a conception would allow us to reconcile the altereity of command with the autonomy of freedom.

All of this should, of course, be spelled out in greater detail, but perhaps enough has been said to indicate at least the direction in which one might move in order to use Heideggerean structures to articulate a non-Heideggerean experience. To get a more complete picture, however, let us move on to a consideration of the Heidegger of the later years. Since we have seen that the problem of freedom is inseparable from the problem of truth, we may safely allow the evolution of the notion of truth to guide us through the turning in Heidegger's way.

II

After *Sein und Zeit*, Heidegger meditated more and more on Being as a process of *aletheia*, and in 1930 he gave for the first time his lecture *On the Essence of Truth*. What strikes him now is this: if Being is the process of *aletheia*, then *lethe* ("-velation," if you will) must somehow antecede the privation of itself, the *a-letheia* (*re*-velation). As a result, Being begins to be conceived now as possessing a certain priority over *Dasein*, a kind of spontaneity by reason of which it reveals itself to *Dasein*. With this experience the so-called "later" Heidegger emerges.

In this new phase, what is to be said of Being? It reveals itself as *Aletheia* in beings and as beings, but because of itself Being is not a being, it hides itself in beings too. As a result, every manifestation of Being is finite, i.e., is constricted within the finite beings that it lets appear. Every revealment, then, is at once a concealment of the rich plenitude of Being, and this phenomenon of simultaneous revealment-concealment Heidegger calls "mystery." In this spontaneous disclosure of itself in beings to *Dasein*, Being is said to "send" (or "e-mit") itself (*sich schikt*), and *Dasein* is at the same time "com-mitted" (*Schicksal*) to the process. This process of e-mitting-com-mitting, taken as a correlation between Being and *Dasein*, is called "mittence" (*Geschick*), which, of course, is always a finite phenomenon. Now what characterizes any given epoch of history is precisely the way Being reveals itself (and conceals itself, too, for of course the mittence is finite) in beings at a given time. In other words, every epoch is determined by a finite mittence of Being. For example, the epoch of Absolute Idealism was characterized by the finite mittence of Being to Hegel; our own epoch is characterized by what Heidegger calls the mittence of "technicity" (*Technik*). At any rate, these epochs (mittences: *Geschick-e*) taken together constitute inter-mittence (*Ge-schick-te*), which is to say history (*Geschichte*), i.e., Being-as-history.

What now of *Dasein*? It is the *Da des Seins*, the There of Being among beings through which Being reveals itself. Being has need of its There, so that the revelation can take place; *Dasein*'s task is simply to let Being reveal itself in the finite mittence, to let Being be. Sometimes the revelation of Being to *Dasein* is conceived as a "call" or "hail" to *Dasein*. *Dasein*'s task is, then, to "respond" to that call, to "correspond" with it, to "tend" Being in beings as the "shepherd" of Being, to acquiesce to its own commitment in the e-vent of Being's self-revelation. It is this acquiescence of *Dasein* to Being-as-revelation that Heidegger now calls "thought"--"foundational" thought.

There can be no question of elaborating here the conception of foundational thought. We must restrict our attention to the question of freedom and its implications for morality. We can situate the problem best if we first see clearly that the question that preoccupies the later Heidegger is no different from the question of Heidegger I: What is the meaning of Being? The difference between the two is simply this: in the early years Heidegger approaches the question through an analysis of *Dasein*; in the later years he tries to think Being for itself and from itself. Our question about freedom, then, comes down to this: How is the conception of freedom, already articulated in *Being and Time*, transformed in the later period and in particular with reference to the nature of foundational thought?

Recall that Being (*Aletheia*), revealing itself in finite mittence, conceals itself as well. This self-concealment (which again is itself concealed in a type of compound concealment) is called "mystery" and is a first type of non-truth (i.e., limitation) intrinsic to truth itself. Another type of non-truth is called "errance" (*Irre*), i.e., the self-concealment involved in *Aletheia* is such that it even beguiles *Dasein* into forgetfulness of the mystery, makes beings themselves seem to be what they are not. Now for *Dasein* to correspond to Being (*Aletheia*) in terms of this double negativity, it must discern Being *(Sein)* from merely seeming-to-be *(Schein)*. This discernment Heidegger calls a "scission" *(Scheidung)*, but just such a scission is a "de-cision" *(Entscheidung)* of thought. Of such a nature was the effort at thought among the Pre-Socratics, such must be the structure of foundational thinking.[18] But this acquiescence to the coming-to-pass of *Aletheia* in all of its negativity--what is this but the gesture of resolve by which, according to *Being and Time*, authenticity is achieved?[19] Indeed! And Heidegger himself is very explicit about the point. "...The essence of thinking [is]... resolve unto the presencing of truth."[20] We infer, then, that it is by foundational thinking that *Dasein* achieves its authenticity and thereby becomes authentically free. Here only the focus has changed. When authenticity is conceived as the result of foundational thinking, there is less emphasis on it as the achieving of the self than upon the aspect of responding to a hail or the accepting of a gift. We will find the same emphasis transposed into a different key in the conception of freedom. Let us see this more in detail.

To begin with, since Being is *Aletheia*, the originating process of revealment-concealment, it is itself by the same token *the Free* (*das Freie*), and each epochal mittence constitutes in its own way the freedom in which *Dasein* finds itself.

Freedom permeates [*verwaltet*] the Free in the sense of something lit-up, i.e., revealed: To the coming-to-pass of revealment, i.e., of truth, freedom stands in the closest and most intimate relationship. [And] all revealing is inseparable from a hiding and concealing. What has been concealed, however, and continues to conceal itself is the Source of all liberation, Being-as-mystery. All revealment comes from the Free, goes toward the Free, and brings [Dasein] into the Free. The freedom of the Free consists neither in the license of the arbitrary nor in restriction by mere laws. Freedom is what conceals [itself] in lighting up [beings]. In this lighting-process there wafts that veil that conceals the process by which all truth comes-to-presence, and [at the same time] lets the veil itself shine forth as doing the concealing. Freedom is the domain of mittence that at any given moment sets revealment on its way.[21]

If Being, then, is the supremely Free, sending itself in finite (i.e., self-concealing) mittence of freedom to man, how conceive the freedom of man? "...Man becomes free for the first time precisely insofar as he becomes an attend-ant of the domain of mittence and thereby someone attent-ive [to its hail]..."[22]--in other words, insofar as he acquiesces to the epochal revelation of *Aletheia*. This revelation is addressed to him as a hail--not imposed upon him as a constraint (*Zwang*) but bestowed as a gift that before all else liberates him unto the fulness of his power. "...Being, insofar as it e-mits itself to man..first liberates men into the Free of the essential potentialities of any given commitment."[23] Thus rendered free, he *can* (freely) respond to the hail.

...The hail brings our essence into the Free, and this in so decisive a manner that what calls us to thought gives [us] the freedom of the Free in order that what is free in a human way can dwell there. The originating essence of freedom con-

ceals itself in the hail that gives to mortals [the task] of
thinking that which above all else is to be thought [i.e.,
Being (Aletheia) itself]....[24]

Briefly, then, Being (Aletheia) for the later Heidegger is
itself the Free, and each of its mittences constitutes a domain of
freedom in which Dasein is first liberated unto the power freely
to accept the gift of Being's revelation. The freedom of Dasein
consists in that gesture of acquiescence to (foundational thought
of) the revelation by accepting its gift with gratitude. In this
sense Heidegger describes this supreme moment of thinking as
thanking (Danken).[25]

If we were to appreciate the full import of this freedom as
Heidegger conceives it, we would have to follow his own analysis
of authentic response to a mittence of Being such as he described
it, for example, in "The Question about Technicity," where he
himself reflects on the mittence that constitutes our own epoch of
Being-as-history, i.e., technicity.[26] But this would take us too
far afield. Instead, let us stop here and attempt to consolidate
our gains by returning to the problem of morality.

Heidegger II is no more concerned with morality than Heidegger I, and he has a chance to articulate his attitude on the
matter very explicitly in the Letter on Humanism, when one of the
three questions that had been posed to him by Jean Beaufret
dealt with the problem of Ethics: "How can one render more precise the relation between ontology and a possible Ethics?"[27]
Ethics, in the sense of a separate philosophical discipline, first
appeared on the scene with Plato, Heidegger claims, when Being
ceased to be experienced as the revealment-concealment of Aletheia, after the manner of the great Pre-Socratics (who spoke of
it rather as phusis), and was considered rather an Idea. Not only
was the genuine sense of Being, then, forgotten, but the original
sense of ethos, too, for this signified to the early thinkers
"sojourn" in the presence of emerging phusis. Thus the tragedies
of Sophocles would articulate a more original meaning of ethos
than is to be found in all of Aristotle's lectures on Ethics.[28] Be
that as it may, we can see how Heidegger situates his own problematic with regard to Ethics as a philosophical discipline:

...If, according to the fundamental meaning of the word
ethos, the name "Ethics" is supposed to say that it meditates
upon the sojourn (Aufenthalt) of man, then that type of
thought which thinks the truth of Being as the originating
element of man [conceived] as an ek-sistent being is in itself

the original Ethics....[29]

In such a perspective we can go even further. If we grant that foundational thinking is "in itself the original Ethics," then we may also say that Being in its mittences is likewise the original Moral Law that Ethics normally meditates.

Only insofar as man, ek-sisting in the truth of Being, is an attend-ant [gehört] of Being, can come the dispensation of those intimations which are to become law and rule for man. To "dispense" in Greek means *nemein*. The *Nomos* is not only law but more originally the dispensation of Being hidden in [its] mittence [to Dasein]. Only this dispensation is capable of meshing man with Being. Only such a mesh can sustain and bind [him]. Otherwise, all law remains no more than the artifact of human reason. More essential than all rule-making is [the fact] that man sojourns in the truth of Being....[30]

For Heidegger II, Being, then, is conceived not only as *Aletheia* but as *Nomos*--and eventually as *Logos*, too. We must be content here merely to indicate the fact and remark that whether as *Aletheia*, or *Nomos*, or *Logos*, Being (the Free) is always mittent in character, i.e., reveals-conceals itself in epochs of history, and the foundational thinking (i.e., original Ethics) in man that responds to Being-(*Nomos*)-as-history is essentially a historical thought (Ethics).

III

Let us now summarize and conclude. We are asking if the thought of Martin Heidegger can help us in our own quest of freedom in an age of Christian renewal. More specifically, can he help in any way to think the problems of morality, especially a "new" morality? We have followed a sinuous path, attempting to trace the essential elements of his conception of freedom. The key to his insight is the realization that freedom is essentially not some power or faculty in man but the process of *Aletheia* which liberates from concealment. In the early years this is identified with the process of transcendence and comes to its fulness by the gesture of resolve through which authenticity is achieved. In the later years, after the focus has shifted from *Dasein* to Being it-self, this process is essentially a gift from Being, conceived now as the Free, to which *Dasein*, already the ek-sistent There of Being (the Free), responds. The response is acquiescence to this mittence in all of its finitude, i.e., to the epochal revelation of *Aletheia* that conceals itself even as it reveals itself, and corresponds to what for Heidegger I was resolve. It is clear that Heidegger is not at all concerned with the problem of morality as such. In both periods he is concerned only with Being and Being-structure. We have already raised the question as to whether or not the ontological structure of Dasein discerned by the phenomenological analyses of the early period might suggest new approaches to the ontology of the moral life. Let us conclude with some questions about the later period.

If the freedom of *Dasein* is the gift of Being (the Free), do we not have a way of reconciling a genuine freedom of *Dasein* with the altereity of its Source? And if this Source is Being-as-dispensation (*Nomos*, Law), then would we not accomplish by the same correlation a reconciliation of the freedom of *Dasein* with its cor-respondence with Law? For Law would be given to Dasein as making claim to be accepted, but given as gift--gift precisely of original freedom to be freely accepted in authentic response. Again, if *Aletheia* (the Free) is not only *Nomos* (Law) but *Logos*, do we not have a new way perhaps of thinking the delicate relationship between Law and conscience? For conscience itself is the existential component called "logos" (*Rede*) in *Dasein*, itself the There of *Logos* (Being), so that Being (*Logos*) would utter its call to *Dasein* through the voice called "logos" in *Dasein*, i.e., its conscience. Furthermore, since *Dasein* always finds itself "thrown" (and, indeed, by Being, whose There it is) into a complex of concrete possibilities which might legitimately be called its "situation," through which the revelation of *Logos* is filtered, would we have the right to conceive of *Logos*-as-Law *(Aletheia)* revealing itself through logos-as-situation in logos-as-conscience, hailing *Dasein* to achieve authenticity in terms always of a particular concrete situation? Would such a perspective help us to articulate a morality that would be validly "situational" without at the same time being utterly Law-less? Again, if Being--*Aletheia*, the Free, Law, *Logos*--reveals itself in mittences that constitute as such epochs of history, then may we find in the preoccupation with the problem of freedom that marks our own epoch, perhaps the sign of a mittence of Being in its own right? If so, then would we find in the Being-structures of Martin Heidegger a way of thinking the ontological dimension, i.e., the dimension of Being-as-history, of a purely ontic phenomenon, i.e., the evolutionary process itself? In that case Heidegger might help us come to grips philosophically with such problems as the historicity of human "nature" as such, of the "law" of man's "nature," indeed of truth itself. What relevance such structures might have in coming to grips philosophically with such a problem as the shifting attitude among Roman Catholics towards birth control (to take but one obvious example) is evident.

With questions such as these we are, of course, way beyond Heidegger and in a realm of experience where he would feel out of place. But after we have tried to be faithful to his experience, we have a right to ask if this experience can help us be faithful to our own, i.e., as Christians. Such a question is our own way of achieving resolve in the presence of *Aletheia* in our own time. For to resolve, Heidegger tells us, means to will-to-know, where "knowing" has the sense he finds in the Greek *techne*, i.e., of standing within the revelation of the Being of beings. To will-to-know in this sense means to question. "...Questioning is the willing-to-know that we have just explained: resolve unto the power to take a stand in the manifestation of beings...."[31] In other words, the very raising of the questions we have posed here is one way of achieving authenticity. And the question itself is quest.

NOTES

1. M. Heidegger, Preface to W.J. Richardson, S.J., *Heidegger: Through Phenomenology to Thought*. The Hague: Nijhoff, 1963, p. *xi*. Here and subsequently in these pages all translations are by the present writer unless otherwise noted.

2. *Ibid.*

3. *Ibid.*, p. xi-xiii.

4. "...Der Ueberstieg zur Welt ist die Freiheit selbst..." (M. Heidegger, *Vom Wesen des Grundes*. Frankfurt: Klostermann, 1955, p. 43; hereafter WG).

5. "...Die Selbstheit des aller Spontaneität schon zugrunde liegenden Selbst liegt aber in der Transzendenz..." (*ibid.*, p. 44; Heidegger italicizes whole).

6. M. Heidegger, *Kant and the Problem of Metaphysics*, trans. J. Churchill. Bloomington: Indiana University Press, 1962), pp. 209-255.

7. Heidegger finds his warrant in Kant's explanation of the word "transcendental": "...I call that knowledge transcendental which concerns itself in general not so much with objects *as with our manner of knowing objects insofar as this must be a priori*

possible..." (I. Kant, *Kritik der reinen Vernunft*, ed. R. Schmidt, Hamburg: Felix Meiner Verlag, 1952, B 25; Kant's italics).

8. "Ein endlich erkennendes Wesen vermag sich zum Seienden, das es selbst nicht ist und das es auch nicht geschaffen hat, nur dann zu verhalten, wenn dieses schon vorhandene Seiende von sich aus begegnen kann. Um jedoch als das Seiende, das es ist, begegnen zu können, muss es im vorhinein schon überhaupt als Seiendes, d.h. hinsichtlich seiner Seinsverfassung 'erkannt' sein...Endliches Wesen bedarf dieses Grundvermögens einer entgegenstehenlassenden Zuwendung-zu... In dieser ursprünglichen Zuwendung hält sich das endliche Wesen überhaupt erst einen Spielraum vor, innerhalb dessen ihm etwas 'korrespondieren' kann. Sich im vorhinein in solchem Spielraum halten, ihn ursprünglich bilden, ist nichts anderes als die Transzendenz, die alles endliche Verhalten zu Seiendem auszeichnet..." (M. Heidegger, *Kant und das Problem der Metaphysik*. Frankfurt: Klostermann, 1950, pp. 69-70.

9. "...Der Mensch ist ein Seiendes, das inmitten von Seiendem ist, so zwar, dass ihm dabei das Seiende, das er nicht ist, und das Seiende, das er selbst ist, zumal immer schon offenbar geworden ist. Diese Seinsart des Menschen nennen wir Existenz..." (*ibid.*, p. 205).

10. See M. Heidegger, *Being and Time*, trans. J. Macquarrie and E. Robinson. London: SCM Press, 1962, pp. 32, 67.

11. "...Der volle existenzial-ontologische Sinn des Satzes: 'Dasein ist in der Wahrheit' ist gleichursprünglich mit: 'Dasein ist in der Unwahrheit'..." (M. Heidegger, *Sein und Zeit*. Tübingen: Niemeyer, 1960, p. 222; hereafter SZ).

12. "...Erschlossen in seinem 'Da,' hält es sich gleichursprünglich in der Wahrheit und Unwahrheit. Das gilt 'eigentlich' gerade von der Erschlossenheit als der eigentlichen Wahrheit. Sie eignet sich die Unwahrheit eigentlich zu..." (SZ, pp. 288-89).

13. "Die Angst offenbart im Dasein das *Sein zum* eigensten Seinkönnen, das heisst das Freisein für die Freiheit des Sich-selbst-wählens und -ergreifens..." (SZ, p. 188).

14. "...Das Dasein ist rufverstehend *hörig seiner eigensten Existenzmöglichkeit*. Es hat sich selbst gewählt" (SZ, p. 287; Heidegger's italics).

15. See *Kant and the Problem of Metaphysics*, pp. 214, 224.

16. "...Dieses wesenhafte Schuldigsein ist gleichursprünglich die existenziale Bedingung der Möglichkeit für das 'moralische' Gute und Böse, das heisst für die Moralität überhaupt und deren

faktisch mögliche Ausformungen. Durch die Moralität kann das ursprüngliche Schuldigsein nicht bestimmt werden, weil sie es für sich selbst schon voraussetzt" (SZ, p. 286).

17. In this context the following text, markedly Kantian in tone, is worth more attention than we can give it here: "...In diesem transzendierenden Sichentgegenhalten des Umwillen geschieht das Dasein im Menschen, so dass er im Wesen seiner Existenz auf sich verplichtet, d.h. ein freies Selbst sein kann..." (WG, p. 43).

18. See M. Heidegger, *An Introduction to Metaphysics*, trans. R. Manheim. New Haven: Yale University Press, 1959, p. 110.

19. See *ibid.*, pp. 111-15.

20. "Dann wäre das Wesen des Denkens, nämlich die Gelassenheit zur Gegnet, die Entschlossenheit zur wesenden Wahrheit" (M. Heidegger, *Gelassenheit*. Pfullingen: Neske, 1959, p. 61).

21. "Die Freiheit verwaltet das Freie im Sinne des Gelichteten, d.h. des Entborgenen. Das Geschehnis des Entbergens, d.h. der Wahrheit, ist es, zu dem die Freiheit in der nächsten und innigsten Verwandtschaft steht. Alles Entbergen gehört in ein Bergen und Verbergen. Verborgen aber ist und immer sich verbergend das Befreiende, das Geheimnis. Alles Entbergen kommt aus dem Freien, geht ins Freie und bringt ins Freie. Die Freiheit des Freien besteht weder in der Ungebundenheit der Willkür, noch in der Bindung durch blosse Gesetze. Die Freiheit ist das lichtend Verbergende, in dessen Lichtung jener Schleier weht, der das Wesende aller Wahrheit verhüllt und den Schleier als den verhüllenden erscheinen läst. Die Freiheit ist der Bereich des Geschickes, das jeweils eine Entbergung auf ihren Weg bringt." (M. Heidegger, *Vorträge und Aufsätze*. Pfullingen: Neske, 1954, pp. 32-33; hereafter VA). Compare *ibid.*, p. 258; *Ueber den humanismus* (Frankfurt: Klostermann, n.d.), p. 30; *Unterwegs zur Sprache* (Pfullingen: Neske, 1959), p. 197.

22. "...Denn der Mensch wird gerade erst frei, insofern er in den Bereich des Geschickes gehört und so ein Hörender wird, nicht aber ein Höriger" (VA, p. 32).

23. "...Weil Sein, indem es sich zuschickt, das Freie des Zeit-Spiel-Raumes erbringt und in einem damit den Menschen erst ins Freie seiner jeweils schicklichen Wesensmöglichkeiten befreit" (M. Heidegger, *Der Satz vom Grund*. Pfullingen: Neske, 1957, p. 158; hereafter SG).

24. "...Das Geheiss bringt unser Wesen ins Freie und dies so entschieden, dass Jenes, was uns in das Denken ruft, aller-

erst Freiheit des Freien gibt, damit menschlich Freies darin wohnen kann. Das anfängliche Wesen der Freiheit verbirgt sich im Geheiss, das den Sterblichen das Bedenklichste zu denken gibt..." (M. Heidegger, *Was heisst Denken?* Tübingen: Niemeyer, 1954, p. 153). See also *ibid.*, p. 97. Compare SG, pp. 44, 157, 158.

25. WD, pp. 85, 93, 94.

26. M. Heidegger, "Die Frage nach der Technik," VA, pp. 13-44. The texts cited in notes 21 and 22 above were taken from this essay.

27. M. Heidegger, *Ueber den Humanismus*, pp. 38-46; hereafter HB. It is impossible here to enter into the treatment of morality in *An Introduction to Metaphysics*, pp. 196-199, although a fuller study than is feasible here would demand a consideration of these pages. For a succinct but comprehensive (and thoroughly competent) résumé of the ethical problem in Heidegger, see the admirable work of Reuben Guilead, *Être et liberté: Une étude sur le dernier Heidegger*. Louvain: Nauwelaerts, 1965, pp. 119-125.

28. HB, p. 38.

29. "Soll nun gemäss der Grundbedeutung des Wortes *ethos* der Name Ethik dies sagen, dass sie den Aufenthalt des Menschen bedenkt, dann ist dasjenige Denken, das die Wahrheit des Seins als das anfängliche Element des Menschen als eines eksistierenden denkt, in sich schon die ursprüngliche Ethik..." (HB, p. 41).

30. "Nur sofern der Mensch, in die Wahrheit des Seins eksistierend, diesem gehört, kann aus dem Sein selbst die Zuweisung derjenigen Weisungen kommen, die für den Menschen Gesetz und Regel werden müssen. Zuweisen heisst griechisch *nemein*. Der *nomos* ist nicht nur Gesetz, sondern ursprünglicher die in der Schickung des Seins geborgene Zuweisung. Nur diese vermag es, den Menschen in das Sein zu verfügen. Nur solche Fügung vermag zu tragen und zu binden. Anders bleibt alles Gesetz nur das Gemächte menschlicher Vernunft..." (HB, pp. 44-45).

31. "...Fragen ist das oben erläuterte Wissen-wollen: die Ent-schlossenheit zum Stehenkönnen in der Offenbarkeit des Seienden..." (M. Heidegger, *Einführung in die Metaphysik*. Tübingen: Niemeyer, 1953, p. 17).

10. MARION HEINZ

THE CONCEPT OF TIME IN HEIDEGGER'S EARLY WORKS*

According to Heidegger's own self-understanding, the philosophy of time which was developed in *Being and Time* and related works,[1] is neither to be classified in the context of the traditional theories of time, nor is it to be understood as a contribution to the analyses of time found in contemporary philosophy of life or in phenomenology. As stated by Heidegger himself, this philosophy of time is concerned mainly with a definition of time that is of a completely new style; its superiority over other theories should prove itself above all by the fact that one can show that the traditional philosophical as well as the everyday understanding of time receives its foundation from it.[2]

According to Heidegger, the traditional theories of time, developed between Aristotle and Bergson, exhibit two common characteristics, all significant differences as far as particulars are concerned notwithstanding:[3] 1) The systematic place for the treatment of the phenomenon of time is philosophy of nature. Time, just as space, is treated in the context of the determination of the unity and the order of natural processes. 2) Time is understood from the now-moment. Its parts, namely future, past, and present, are conceived as a now that is not yet, a now that is no more, or a now that is. Time itself is nothing but the succession of these now-moments. The philosophical conception of time shares this view with the everyday understanding of time. For the latter, time is the "frame" into which the historical processes, natural events, and one's own life enter. Time is portrayed as a stream of now-moments which arrive from the future and flow away into the past.[4]

It is not Heidegger's intention to prove that these explanations of time are wrong and to replace them with his own

---

*Translated for this volume by Joseph J. Kockelmans.

correct theory. By demonstrating that the traditional and everyday determinations of time are to be founded on a more original
concept of time, Heidegger explicitly asserts that these conceptions can be justified, even though only within limits.[5]

Within the history of the theories of time, which he interprets as being homogeneous in principle, Heidegger sees in Kant
an exception and portrays him as an advocate of his own
approach.[6] Kant has attributed to time a function that is fundamentally different from that given to it by the tradition. In
Heidegger's interpretation of Kant's philosophy, time is in essence
self-affection;[7] taken as such it constitutes the basic structure of
the subject in such a way that through the production of horizonal schemata it renders at once also the objectivity of the objects
possible. In this respect Heidegger finds in Kant an approach
predelineated which is in correspondence with his own intention:
the foundation of the possibility of ontology on the basis of time
taken as the basic structure which first renders possible both
subjectivity and objectivity.

STARTING POINT, METHOD, AND STRUCTURE OF THE INVESTIGATION CONCERNING TIME

The basic task of *Being and Time* is to work out the question concerning the meaning of Being[8] and in this way to give a
foundation to ontology as "a non-deductive genealogy of the different possible ways of Being,"[9] and the thematization of the
possible unity of the concept of Being. It is from the perspective
of this basic task that time becomes the central theme in *Being
and Time*. According to Heidegger the question concerning the
meaning of Being transcends by far the manner of questioning
found in the traditional Aristotelico-Thomistic doctrine of Being,
insofar as in his ontology not only the being as being is inquired
into with respect to its essential determination, but also that
which determines the beings, namely Being itself, is thematized in
its own determination. "However, in order to be able to understand the essential determination of this being through Being, the
determining element itself must be understood with sufficient
clarity. It is necessary, therefore, first to comprehend Being as
such, and this comprehension must precede that of the beings as
such. Thus the question *ti to on* implies a more original question:
what is the meaning of Being which is preunderstood in this
question?"[10]

The basic thesis which Heidegger pursues in *Being and Time* and related works can be formulated as follows: the meaning of Being is time.[11] The manifold modifications of Being as well as the uniform determination of Being as distinct from the beings become understandable from the perspective of time.[12] However, the traditional conceptions of time are in principle inadequate for the interpretation of time as the horizon for our comprehension of Being, insofar as in those conceptions time itself is preunderstood as a being which is to be determined in its Being.[13] Thus time becomes "questionable in the same way as Being."[14]

The first step in the elaboration of the question of Being consists in the analysis of the structures of our understanding of Being which always is already factical. Now insofar as our understanding of Being is a constitutive moment of Dasein's own mode of Being, the analytic of Dasein is fundamental ontology.[15] Under the hypothetically accepted premise that Being fundamentally can be understood only from the perspective of time, the constitution of Dasein's Being must itself receive a temporal meaning. The preliminary aim of *Being and Time* is the exposition of this temporal meaning of the Being of Dasein, which is signified with the term "temporality" (*Zeitlichkeit*).[16]

From a methodical perspective the analytic of Dasein proceeds hermeneutically, i.e., it retraces our original, existentiell understanding by interpreting it, in order to articulate it conceptually with respect to its structures and the ontically hidden conditions of its possibility.[17] Insofar as that which is thematized in this way consists in the modes of revealment, or the modes of non-concealment, of the beings and of Being, the method is phenomenological, i.e., the showing and determining of the modes in which Being and the beings show themselves.[18] The hermeneutic-phenomenological method predelineates the further steps in the explanation of the question of Being.

Interpretative explanation proceeds in a circular manner: something that is understood already is explicitly projected upon that, upon which it was already projected in the original understanding, in order that one can now explain and comprehend it *as* something from this "upon-which," taken as the ground of its being-understood, i.e., its meaning.[19] Insofar as this explaining projection in principle is subject to the possibility of error, it is necessary to ascertain its adequacy. The ground of our understanding which has been made explicit in the explanation (i.e., that upon-which on the basis of which something becomes understandable as something) must in a second methodical step, namely the reinterpretation of what on the basis of this ground has been understood, be verified *as* ground.[20] The sequence of these methodical steps, namely "basic explanation" and "reinterpretation," is determinative for the methodical procedure of *Being and Time* in such a way that the basic explanations, in addition, show a gradual succession of steps with respect to their originality.

A further methodical directive follows from the special character of the Being of Dasein that is made the theme of fundamental ontology: the Being of Dasein taken as ek-sistence is essentially accomplishment, i.e., the eksistential concepts, in contradistinction to the categorial determinations of something that is merely present-at-hand, are nothing but concepts of modes of accomplishment. The ontological determinations of Dasein, therefore, must phenomenally be shown to be ontic modes of accomplishment.[21]

These methodical requirements determine the scheme according to which Heidegger's analyses of time are developed: temporality taken as the ground of the unity of care (i.e., the ontological concept of the Being of Dasein) is revealed from the perspective of the authentically materialized Being of Dasein, i.e., of the anticipating disclosedness (cf. § 65). The temporal reinterpretation of Being-in (§ 68) verifies temporality as the ground of care. The analysis of authentic historicity (§§ 73-75) and the analysis of the "ordinary" and traditional comprehension of time (§§ 79-81) describe the exhibition of the existentiell, authentic or inauthentic, understanding of temporality.[22] The derivation of "ordinary time" is meant to justify that temporality is the original phenomenon of time.[23] The transition from fundamental ontology, taken as the working out of the meaning of the Being of Dasein, to the temporal ontology taken as the working out of the meaning of Being in general, has not been materialized in *Being and Time*.[24] However, from the published lectures, and especially from the lecture of the summer semester of 1927, *The Basic Problems of Phenomenology*,[25] one can infer that this transition was to be derived from the elucidation of the temporal meaning of Being-in-the-world (cf. BT, § 69c). But the last methodical step, namely the reinterpretation of the analysis of Dasein from the elucidated meaning of Being in general,[26] is meanwhile, not even in a schematic form, treated in the texts which thus far have been published; this is thus contrary to what has been done with respect to the elaboration of the temporality of Being.

THE TEMPORALITY OF DASEIN

According to Heidegger, the phenomenal content which materializes (*erfüllt*) the signification of the term temporality, is drawn from the structure of Being of anticipatory resoluteness.[27] This mode of Being presents the existentiell mode of the authentic Being-a-whole; i.e., Dasein understands itself as a whole in the extreme and unsurpassable limits of its Being-disclosed, qua death and thrownness, as a being that has to determine itself with respect to concrete and factical possibilities of its Being-in-the-world.[28] The unity of the structural moments of its disclosedness, anticipating death, understanding one's own Being-guilty, and "bringing oneself in the situation of action" can be made understandable, Heidegger feels, only when these structural moments themselves are interpreted as phenomena of time.[29]

By anticipating death, Dasein makes the possibility its own as which it always is ahead of itself as long as it is, namely to be the individualized and not to be outstripped Being-in-the-world as self. The meaning of such an understanding is for Heidegger the future.[30] But future cannot mean here a present-that-is-not-yet, insofar as Dasein precisely *is* as anticipating the possibility of its most authentic Being-in-the world. Formulated paradoxically one could say that with the term "future" Heidegger conceives of a becoming that constitutes Being: Dasein is authentically in such a way that it becomes (understands, grasps), what it already is in the manner of Being-ahead-of-itself. The formal structure of the future, which is still indifferent in regard to authenticity and inauthenticity, has been illustrated by Heidegger in the following manner:

"(The question mark indicates the horizon that remains open.")[31] Dasein's Being-futural is made explicit phenomenally from the anticipation of death as "movement," taken in the sense of Being-toward-possibilities and in the sense of letting oneself come to oneself in such possibilities.

The second moment which constitutes the anticipating disclosedness, is the understanding of Being-guilty. Dasein understands that it does not eksist *through* itself as a being of this mode of Being, but rather that it finds itself being present as a being, and has to take charge of itself in that mode.[32] According to Heidegger such an understanding is grounded in the temporal phenomenon of having-been-ness.[33] Just as little as was the case with the future, so having-been-ness, too, cannot be grasped from the perspective of the "ordinary" understanding of the past as a present-that-no-longer is. Understanding Dasein is its thrownness, and the latter does not lie behind it as something that is just bygone.[34]

Having-been-ness for Heidegger has the structure of a coming-back-to itself.[35] By coming toward itself from its ownmost possibility Dasein understands that in each case it was already this very possibility.

According to Heidegger, the "situation," taken as the whole of factical eksistence which in resolve has been chosen in favor of determinate possibilities of concern and solicitude, is grounded in the present. This phenomenon of time has the character of something that lets innerworldly being be encountered.[36]

Summarizing we get the following result: each moment of the anticipating resolve is grounded in a phenomenon of Dasein's temporality. The character of these moments of original time is determined through the moments of the mode of Being of anticipating resoluteness in the following manner: "the future, having-been-ness, and the present, show the phenomenal characteristics of the 'towards-itself,' the 'back-to,' and the 'letting-oneself-be-encountered by'."[37]

In order to make Heidegger's thesis that temporality is the ground of the unity of the structural whole of anticipating resoluteness, or of care as such, understandably a more careful determination of temporality is necessary. According to Heidegger, original temporality should not be represented as a one-dimensional continuum with the help of the image of a line, but rather, to a certain extent as a three-dimensional "extendedness." The "dimensions" of temporality, namely future, having-been-ness, and present, are conceived here ekstatically and horizonally; they "are" in the manner of the equiprimordial temporalizing.[38]

The term "ekstasis" characterizes the temporal phenomena formally as genuinely different modes of standing-out, of carrying-away towards a whereto. Taken according to its content, temporality is grasped in this way as a "Being-open for..." that is oriented in different directions. "Every such carrying-away is intrinsically *open*. A peculiar *openness*..., belongs to ekstasis."[39] This Being-open is thought here as something that opens itself and not as something open that is simply present-at-hand. "The ekstasy mentioned here, this stepping outside itself (*ekstasis*) is to some extent a *raptus* (rapture). This means that Dasein does not become gradually expectant by traversing serially the beings that factually approach it as things in the future, but this traversing rather goes gradually through the open path made way by the *raptus* of temporality itself."[40]

The open that has been formed in the ekstatic carrying-away becomes determined more carefully through the concept of extendedness as "stretchedness"[41] that is without gaps and breaches so that temporality forms a "steadiness that has been stretched out."[42]

The ekstatical "Being-open for..." is adequately characterized through this horizonal structure. The ekstases of temporality are not simply "modes of being carried away to..., not a being carried away, as it were, to the nothing. Rather, as a mode of being carried away to..., and thus because of the eksatic character of each of them, they each have a *horizon* which is prescribed by the mode of the *raptus*, the carrying-away, the mode of the future, past, and present, and which belongs to the ekstasis itself."[43] The ekstases are characterized by the horizonal structure as limited, determined modes of the Being-open such that the horizons taken as the boundaries of the ekstatic *raptus* which themselves remain unthematic, predelineate the modes of the disclosedness of being and Being which in them are possible. Starting with his Kant interpretations, Heidegger conceives of this paradigmatic character of the whereto of the temporal modes of being carried away with the concept "schema."[44] This concept is to express that to the ekstases there belong general predelineations of the manner in which something manifests itself in modes of revealment which are primarily ekstatically founded. Corresponding to the three ekstases the following horizonal schemata are indicated: "The schema in which Dasein comes towards itself *futurally*, whether authentically or inauthentically, is the *for-the-sake-of-itself*. The schema in which Dasein is disclosed to itself in a state-of-mind as thrown, is to be taken as that *in the face of which* it has been thrown and that to which it has been abandoned. This characterizes the horizonal schema of having-been-ness....The horizonal schema for the present is defined by the *'in-order-to'*."[45]

Heidegger asserts that original time, unlike the everyday and traditional understanding which attributes to time the character of infinity, is finite.[46] The phenomenal character of temporality manifests itself primarily in Dasein's running forward towards death: when Dasein is disclosed therein in its extreme and most proper possibility, one can also say that this original Being-futural brings the being-able-to-be to conclusion such that Dasein first understands itself as the finite scope of projection which is defined from the perspective of this extreme possibility.[47]

The "Being" of temporality is the equiprimordial temporalization. This tautological formulation is to express negatively that temporality cannot be conceived of as a being that can be determined from other beings. This expression thus refutes the possible false representations according to which the moments of temporality would be present-at-hand together, would originate the one after the other, or could be reduced to each other.[48] Positively, the unity is rather characterized as the "happening" of the opening of the open, of the clearing, in such a way that this happening is to be presented as the "movement" of a stretching-itself-out of each ekstasis into the others and towards their horizons, which in each case flows from one ekstasis.[49] Temporality, which in the concept of temporalization is thought as "the unity which unifies itself,"[50] functions as "that which primarily regulates the possible unity of all essential eksistential structures of Dasein."[50] The equiprimordial temporalization of the three ekstases which in each case are primarily constitutive for a moment of care, renders the unity of care possible as a totality of moments that cannot be reduced to something else.[52] This means that every actualization (*Vollzug*) of ek-sistence is care and, according to Heidegger, this means that it is constituted equiprimordially by the structural moments understanding, moodness, *logos*, and concern. According to the ekstasis from which temporality temporalizes itself primarily, care stands in the mode of understanding, moodness, or *logos*, etc. The relationship of these modes to care as such is not a relation of species to genus.[53] The plurality of the modes is related to the unity of care not in such a way that these modes would be distinguished from the unity through additional determinations; rather it is the stress within the structural whole, according to which in each case one of the ekstases has a privileged position within temporality, that functions as the differentiating factor. The differences of the modes are merely shifts of this accent or stress within the multiplicity of the moments which for the rest are identical. The following survey shows the manner in which the ekstases and moments of care, or Being-in, are coordinated with each other:

Temporality	Care	Being-in
Future	Ahead-of-itself	Understanding (*Logos*)
Having-been-ness	Already-Being-in	Moodness
Present	Being-with (Fallenness)	Concern (*Logos*)

One has often criticized the fact that there is an incongruency between the moments of temporality and those of care such that *logos* has no place in the structure of temporality.[54] For contrary to what is the case with the other modes of disclosedness, no proper ekstasis is coordinated with *logos*. Even though it is the case that apparently only a three-membered structure of time corresponds to a four-membered structure of the Being of Dasein, this nonetheless does not mean that *logos* would altogether lack a foundation in temporality. For the temporalization of the meaning of *logos*, taken as the articulation of the meaning of Being in general which comes-to-pass in understanding (future), moodness (having-been-ness), and comportment with beings (present), is nothing else but the temporalization of the entire temporality taken as a whole. The scheme of *logos* is the "as"[55] which, depending on the mode of temporalization, modifies itself into the apophantic or the hermeneutic "as."[56]

Factically, Dasein does not at all eksist in the basic mode of understanding as such, moodness as such, etc., but rather for instance, in the mode of fear, anxiety, or the pale lack of mood. These modes are temporally interpreted by Heidegger from the possibility of modification of each ekstasis as such.[57] For instance, in the mode of fear having-been-ness becomes modified into "confused forgetting;" the ekstasis which functions as starting point on its part modifies then the other ekstases, so that a correspondence of the ekstases is formed, for instance with respect to the modes of their being-confused.[58] Compared with the basic modes such as understanding, moodness, etc., those modes such as fear, curiosity, etc., are more specific phenomena which are characterized through additional determinations.

The relation between fallenness and being-with, or respectively that between authenticity and inauthenticity, is also disputed in the Heidegger literature.[59] The following questions are to be raised here: 1) How can one understand from temporality itself that fallenness is co-disclosed in every mode of care without Dasein having to eksist in each case as falling? Is it perhaps necessary to distinguish between an authentic and an inauthentic falling? 2) How is one to combine the thesis that Dasein eksistentielly comes to authenticity only via the detour of inauthenticity, with the thesis that on the level of temporality the inauthentic temporalization is subordinated to original and authentic temporalization? Does it not follow from this that also on the eksistentiell level authentic eksistence must come first?

Fundamentally the authentic actualization of eksistence is temporalized out of the future, whereas the inauthentic is temporalized out of the present. Thus for authentic temporality the following order among the ekstases is constitutive: "Primordial and authentic temporality temporalizes itself in terms of the authentic future and it does so in such a way that in having-been futurally, it first of all awakens the present."[60] The relation of future and having-been-ness must rather be thought such that that as which Dasein has been, is determined out of the future. The "having-been 'is' in each case only according to the mode of the temporalization of the future, and only in it."[61] The future is extended immediately and without any breaches into having-been-ness.[62] The priority of the future and its determining stretching-itself-out over the whole of having-been-ness constitute the temporal foundation for the possibility of a being that has the mode of Being of eksisting. For if the distinguishing factor of this being consists in that all determinations of its Being are possibilities of itself, then everything which Dasein was already, must be included completely in this Being-possible and, thus, cannot already be established in its qualitative determinations from the outset.

In the authentic temporalization of temporality the present is "included" in future and having-been-ness.[63] At first this must be understood in such a way that the present, as that which lets innerworldly beings be encountered, in its horizon of the "in order to" already presupposes both future and having-been-ness, out of which Dasein is disclosed as that "for" for which the letting-be-encountered takes place. Dasein, as a being that is already, determines an in-order-to from a for-the-sake-of-which. Because the present is the disclosing of the Being of that being which Dasein itself is not, the present is, if taken as such, inauthentic, i.e., as far as the Being of Dasein is concerned it is closing off and concealing.[64] Heidegger's claim that the present is "included" or "held" in authentic temporalization means that the tendency on the part of the present as such to conceal the Being of Dasein is here arrested. As far as content is concerned, this must be understood such that Dasein itself renders possible for itself out of its most proper factical being-able-to-be the possibilities of solicitude and concern, and not conversely that it lets its being-able-to-be (Seinkönnen) be determined from these. Terminologically authentic present is defined as moment of vision (Augenblick). "The moment of vision (Augenblick) is nothing but the look of resolve, in which the full situation of action opens itself and is held open."[65]

For the clarification of the problems mentioned the distinction between truth and certainty must be taken into consideration. Dasein eksists authentically, if it holds itself in the truth of its being; eksistentiell authenticity is a mode of being-certain, of holding-to-be-true. Only when Dasein appropriates to itself that which in an understanding way is disclosed, thus explaining itself as a being from this Being, does it eksist authentically as a being.[66] Interpreted temporally, the interpretation of a being in its Being presents itself as the reversal of the primary understanding disclosure: the present adapts itself to the Being that is disclosed in future and having-been-ness, and brings in this manner the being into view *as* something, i.e., in its Being.[67]

It is therefore possible that Dasein is disclosed authentically, although existentielly it is unable to hold itself in what is so disclosed and, thus, eksists in the mode of inauthenticity. For instance, in anxiety one's own futile (nichtig) being-able-to-be is disclosed in such a way that Dasein at first does not hold itself as a being in this Being, does not come to its self in it, but instead fleeing from this Being loses itself in the submission to beings which in making present is revealed.[68] Thus, it is not only not a contradiction that we find together the priority of authenticity on the level of temporality with the priority of inauthenticity on the eksistentiell level; for only in this manner is it possible to understand our falling to the world in its meaning as a movement of flight.

In the horizon of the present taken as in-order-to, the Being of innerworldly beings becomes disclosed. In view of the fact that consequently in the present the Being of Dasein is disclosed only with respect to its dependence on beings which have this mode of Being, the adaptation of the present to the original Being of Dasein that is disclosed in future and having-been-ness, an adaptation which is required for the interpretation of Dasein as a being in its Being, is basically affected negatively. For of the present the tendency is characteristic to "escape" and to temporalize itself out of itself, such that the Being disclosed in it becomes the dominating meaning of the being.[69] Insofar as making-present closes Dasein's Being off and, thus, is authentic, this making-present is to be considered, in every mode of the temporalization, as the condition of the possibility of falling. Only when temporality temporalizes itself out of the present, is it the case that through its inauthenticity all ekstases become moved over into inauthentic modes so that falling can appear not just as a moment, but rather as the entire unity of care. The being-held (*Gehaltenheit*) of the present in authentic temporality consequently means just a counter-movement; it is thus to be understood as the arresting of this "escaping."

THE TEMPORALITY OF BEING

In the preceding we have shown temporality to be the meaning of the Being of Dasein. This constitutes the foundation of the temporal ontology which in *Being and Time* was conceived, but not carried through. The transition from fundamental ontology to ontology as such was presented in a provisional manner in the course *The Basic Problems of Phenomenology*, which was delivered in Marburg during the summer semester of 1927.

For Heidegger, ontology originates as a science through the "objectivation of Being as such."[70] This objectivation has "the function of *explicitly* projecting what is antecedently given upon that on which it has *already* been projected in pre-scientific experience or understanding. If Being is to become objectified--if the understanding of Being is to be possible as a science in the sense of ontology--if there is to be philosophy at all, then that upon which the understanding of Being, qua understanding, has already pre-conceptually projected Being must become unveiled in an explicit projection."[71]

But if ontology consists in the interpretation of the pre-scientific comprehension of Being, and if furthermore temporality constitutes the condition of a being that has the mode of Being of understanding Being as such, then it must also be the case that one can derive from temporality that "from which we understand something like Being."[71]

Heidegger's thesis can be formulated as follows: Being is understood from the horizonal schemata of temporality. The temporal meaning of Being, namely Temporality (*Temporalität*) is "temporality taken with reference to the unity of the horizonal schemata which belong to it."[73] In the explication of this thesis the starting point for the solution of the basic problem of ontology, namely the distinction between Being and being, is developed at the same time.

The starting point here is the determination of the relationship between intentionality and transcendence as the basic attitudes of Dasein, namely that it is to orient itself towards being and to understand Being. Intentionality is the structure which constitutes the relational character of Dasein's comportment.[74] It is characterized by the following structural moments: 1) *intendere*, to orient itself toward; 2) the *intentum*, that toward which the orienting-itself is oriented, namely the being; 3) a specific directional meaning of the *intentio*, according to which a determinate mode of the discoveredness of the *intentum* is rendered possible. For instance, it belongs to the intentional, directional meaning of perception "to intend the perceived thing as something that is in itself present-at-hand."[75]

According to Heidegger, the possibility that in each intentional comportment toward beings a determinate mode of discoveredness of the beings is already given in advance, presupposes a previous and preliminary understanding of the Being of the *intentum*.[76] Heidegger conceives of this understanding of Being which gives a foundation to the possibility of intentionality, as transcendence.[77] Dasein always has already transcended the totality of beings. That toward which this totality has been transcended, is the world.[78] The understanding of world, which is the understanding of Being that belongs to intentionality, implies the Being of Dasein as well as the Being of the beings which do not have the mode of Being of Dasein.[79]

According to Heidegger the unity of intentionality and transcendence, of comportment toward beings and understanding of Being, is grounded in the unity of the ekstatic and horizonal structure of temporality. The unity of these structural moments of temporality is determined as "self projection."[80] For temporality an "inner productivity" is characteristic[81] in the sense that the ekstases project themselves upon themselves in such a way that the product of the self-projection consists in the horizonal schemata, out of which the ekstases, taken as determinate modes of the "Being-open for...," first manifest themselves.[82] The horizonal schemata which in the self-projection of the ekstases are produced function in a twofold manner as "model" (*Vorbild*); they determine at the same time the specific mode of Dasein's directedness (*Ausgerichtetsein*) and the mode or manner in which the Being of the beings that are encountered therein is understood.[83] On the basis of the horizonal schemata of temporality the way or

manner of this orienting-itself toward beings taken as an attitude
of Dasein, and the manner of the possible Being-revealed of the
beings that have been encountered therein, are in each case predelineated and attuned in regard to one another. By way of
example Heidegger has described this for the present and its horizonal schema, the *praesens (Präsenz)*:[84] for perception the
ekstasis of the present is primarily constitutive. Perception lets
something be encountered in its character of being something-in-
itself. By taking his starting point from this fact Heidegger then
explains this as follows: "An understanding of Being can already
be present in intentional perception because the temporalizing of
the ekstasis as such, enpresenting as such, understands in its
own horizon, thus by way of *praesens*, that which it enpresents,
understanding it as something present *(Anwesendes)*. Put otherwise, a directional sense can be present in the intentionality of
perception only if perception's direction understands itself by way
of the horizon of the temporal mode that makes possible perceiving as such: the horizon of *praesens*."[85]

The horizonal schema of the *praesens* thus "regulates"[86] at
once the intentional comportment of Dasein and the kind of the
discoveredness of the being that has been encountered therein.
Making-present, which is primarily founding for perception, is to
be understood from the horizonal schema of the *praesens*; this is
the reason why it receives the directional meaning which distin-
guishes it as this particular kind of intentional revealing from
other such kinds, namely that it lets the being be encountered in
its bodily in-itself and not, for example, in the manner in which
this is done in concernful occupation, namely by discovering the
being in its reference to and dependence upon an other being.
Together with this we also find the predelineation of that toward
which, and of the manner in which, it lets the being be encountered in this way of revealing.

But if temporality in this way renders possible the preontological understanding of Being which at first is not yet differentiated, then it also gives a foundation to the possibility of ontology
as the objectification of this preontological understanding of Being.[87] The manner in which this objectification proceeds can be
described in the following way: In order to articulate conceptually
Being in its temporal meaning, one must take one's point of
departure from intentional attitudes or approaches. These must
then be determined with respect to their temporal meaning. In
harmony with the ekstases of temporality which are constitutive
for the specific kind of intentionality, the comprehension of Being

can then be explicated from the horizonal schemata of this mode of temporalization. From this it becomes evident that Temporality is not posited as an a priori principle of deduction for the diverse differentiations of Being; the temporal meaning of Being can in each case only be revealed by starting from the factically concrete modes of eksistence.[88]

The temporal ontology whose possibility is grounded by temporality and which here was described in its mode of proceeding only in outline, must solve the following problems: "first, the problem of the *ontological difference*, the distinction between Being and beings; secondly, the problem of the *basic articulation of Being*, the qualitative content of a being and its mode of Being; thirdly the problem of the *possible modifications of Being* and of the *unity of the concept of Being* in its ambiguity; fourthly, the problem of *the truth-character of Being*."[89]

For Heidegger the most important problem of ontology is that of the ontological difference, because the clarification of this distinction decides about the possibility of philosophy, as the thematization of Being, in distinction from the other sciences which thematize the beings.[90]

Not only the explanations concerning temporality as the ground of the possibility of the unity of intentionality and transcendence, but also Heidegger's elucidations about Kant's thesis that "Being is not a real predicate, or that Being is identical with positing, and Dasein with absolute positing,"[91] serve to work out the "principle" of the ontological difference. According to Heidegger, Being basically is not a being's determination that is; it is not the totality of its determinations, either; Being is rather the predelineation of the kind of Dasein's intentional self-realization and of the manner in which the being becomes manifest therein; this predelineation is determined from the horizonal schemata.[92] In this claim the basic idea is implied that Dasein, while it understands itself as a specific mode of the "letting something be encountered," at the same time also understands the specific way in which beings can be revealed. This is the reason why Being always is "something" from which Dasein understands itself, its modes of self-realization, and out of which, at the same time, being becomes understandable in the manner of its non-concealment.[93]

Temporal ontology is fundamentally different from the traditional conceptions of ontology because of this determination of Being. The issue no longer is about the knowledge of the basic determinations of beings, regardless of whether they precritically are understood as things in themselves, or critically as appearances. In Heidegger's case, what is thematized is nothing but the possibility and the manifold modes of the non-concealment of the beings. On the basis of this problematic our understanding of the scientificity of ontology changes. If the main issue in ontology consists in trying to understand the manifold modes of the meaning of Being, then the ontological meaning of presence-at-hand, which according to Heidegger in traditional ontologies was exclusively the dominant one, can no longer stamp the manner in which ontology understands itself.[94] In ontology Heidegger is not concerned with gaining access to unhistorical, apodictically certain kinds of knowledge about an object domain which is just different, namely meaning.[95] In view of the fact that philosophy is constituted basically as the interpretative explanation of the factically historical, preontological understanding of Being, its own truth must be historical, also.[96] The temporal determinations of Being do not lay claim to have the meaning of supratemporal models for all articulations of meaning; rather they describe Heidegger's own self-understanding according to the explicit appropriation of the "truth" of Dasein's understanding of self and world, which was worked out by starting from a determinate, historical situation and was determined by it.

THE "ORDINARY" CONCEPTION OF TIME

The assertions about temporality as the origin of the ordinary and the traditional, philosophical concept of time[97] have a manifold function in the "taxonomy" of *Being and Time*: 1) They serve the function to justify the claim that temporality, which was grasped in an ekstatic and horizonal manner, constitutes the original phenomenon of time.[98] 2) In harmony with the methodical requirement that all eksistential concepts must be shown to be modes in which Dasein can be, temporality cannot be posited as the ground of Dasein which is just present-at-hand, but must be shown to be the ground which in Dasein is disclosed authentically or inauthentically. The construction of historicity shows how temporality is disclosed in Dasein insofar as it has achieved authenticity, whereas the ordinary comprehension of time constitutes the manner in which Dasein inauthentically understands temporality. Indirectly the derivation of the "ordinary" understanding of time serves to "explain" the fact that Dasein erroneously attempts to conceive of its identity in terms of substantiality or subjectivity.[98] The analysis of historicity, on the other hand,

tries to show that the self of Being-in-the-world is to be understood as a happening in which Dasein identifies itself as a being
from the perspective of its finite Being. 3) The mode of eksistence which is characteristic of everydayness, from which the
eksistential analysis took its point of departure, is then reinterpreted in its temporal Being.[100] 4) Last but not least, these
analyses are to restore to the everyday interpretation of time its
right, i.e., they must try to correct the impression as if time
would be exclusively and nothing but temporality, in the sense
made explicit thus far.[101]

Thus it is not Heidegger's intention to claim that temporality
is the "correct" concept of time; rather it is his intention to show
phenomenologically that "time" legitimately can be interpreted and
conceptually articulated in more than one way; yet this is to be
done in such a way that within these modes of understanding one
can distinguish some hierarchy in regard to their originality. Heidegger's thesis is the following: "The ordinary concept of time
owes its origin to a way in which primordial time has been leveled
off."[102]

The justification of this thesis is accomplished in three
steps: The ascertainment that before all measuring of time Dasein
factically comports itself towards time constitutes the point of
departure.[103] From this point the concept of world-time is then
developed,[104] in which in Heidegger's view the Being-in-time of
innerwordly beings is grounded. The eksistential-ontological
interpretation of the use of a clock and of the measuring of time
is the fundament from which Heidegger makes understandable that
the "world-time that in the use of a clock is 'envisioned,'"[105] is
nothing but the "now-time," which is the only time which everyday and fallen Dasein knows and characterizes as an endless,
passing, and irreversible sequence of nows.[106]

In order to determine the time of everyday concern phenomenologically with respect to its structures, Heidegger takes his
point of departure from the phenomena of man's daily comportment
with time. These phenomena are found in part in language which,
as the enunciatedness of Being-in-the-world, bears witness to a
determinate understanding of Dasein; on the other hand, Heidegger takes as a basis also familiar states of affairs such as, for
example, the fact that in everyday life time, as it were, is
experienced as full of holes and thus not as continuous.

The first phenomenal finding is that in the being that concernfully occupies itself with things that are ready-to-hand,
always already a "connection" between time and the thing of

concern is discovered and expressed in the modes of the "now, that...," "then, when...," "on that occasion, when..."[107] This structure of the concernful understanding of time, taken as "datability," is interpreted to mean a "reflexion of the ekstatical constitution of temporality."[108] The ekstasis of the present which is primarily constitutive for concern, taken as "letting innerwordly being be encountered," is interpreted in language as a temporal mode of the relation of Dasein to beings.[109] The specific character of the present also dominates the interpretation of the other ekstases, so that future and having-been-ness just as making-present, are explained as temporal relations to beings.

With the structure of datability each "now," "then," "at that time," also has already a "spanned character with the width of the span varying: 'now'--in the intermission, while one is eating, in the evening, in the summer; 'then'--at breakfast, when one is taking a climb, and so forth."[110] This second structural moment of the time of concern is founded on the fact that Dasein interprets itself as the ekstatic "extended-ness" of temporality.

The time which is expressed as "now," "then," and "at that time," is always already made public.[111] This means first that the different datings are also "commonly" understood from the environmental occurrences on the basis of Being-together-with-one-another-in-the-world. The "genuine" making-public however comes about in the measurement of astronomical and calendar time. When Dasein dates and divides time from the perspective of the being that is accessible to "everybody," namely the course of the sun, the clock, albeit still as a natural one, is already discovered, i.e., a universal indication of time and a public measurement of time are then given as well.[112]

Finally, in concern time shows itself according to the structure of suitability or unsuitability. Every "now," "then," and "at that time" is related to a for-what; for example, in the expression: "When it gets light, it is time for our daily work." Time itself is accordingly, just as are the inner-worldly beings, understood from the context of the relations of a for-the-sake-of or an in-order-to, i.e., from the total-meaningfulness; thus it has world character. This is the reason why Heidegger calls it world-time.[113] World-time makes the within-time-ness of the beings possible.[114]

For the genesis of the "ordinary" understanding of time from the perspective of world-time, the manner in which we relate to time when we make use of a clock, is decisive. The specific element in the use of a clock consists in this that the dating is

accomplished not from something ready-to-hand which we encounter in our environment, but rather from something that is
present-at-hand.[115] This dating is a "making-present of something that is present-at-hand;" insofar as it is a relation to a
being that is present-at-hand, dating has the character of being
a measurement. A regular occurrence, for instance the motion of
the clock's hand over a determinate distance in space, must, in
order to be able to function as the standard for the measurement
of time, be counted. For Heidegger what is ontologically decisive
consists in this that the being from which the dating of the now
follows, is encountered in its being-present as something that is
present-at-hand. Insofar as time, with respect to its "how much
time there is," must be determined through the numbering of the
thing present-at-hand, time shows itself in the employment of a
clock as "that which is *counted* and which shows itself when one
follows the traveling pointer, counting and making present..."[116]
According to Heidegger, this definition of time which has been
discovered here phenomenologically is nothing but Aristotle's
definition of time: *"touto gar estin ho chronos, arithmos kineseos,
kata to proteron kai husteron.*[117] For this is time, that which is
counted in the motion which we encounter within the horizon of
the earlier and later."[118] It is characteristic for this understanding of time, which is accomplished within the horizon of the
making-present of a thing present-at-hand, that one no longer
sees world-time which one has to measure by means of the
standard of something that is present-at-hand, but only still sees
the nows as the making present which has been expressed while
one was following the pointer.[119] In this way Dasein understands,
in harmony with the meaning of the making present of something
present-at-hand, its temporality itself as something present-at-
hand, namely as the sequence of nows that is present-at-hand.
These nows have been reduced to a sequence of mere points and
they no longer show the relations, which are characteristic for
them in our making-present of something ready-to-hand, namely
their datability, stretchedness, publicness, and its worldly character.[120] However, this leveling off is not just accidental;
according to Heidegger, ultimately it is the expression of the
falling shunning of death.[121] If Dasein turns its attention to its
death, it no longer can see time precisely as such an intersubjec-
tively given sequence of nows; rather it then understands time as
its own, of which it must make use in harmony with the possibilities of its own Being. Only the irreversibility of the sequence of
nows still mirrors the *origin* of the everyday understanding of

time, namely the original and finite temporality.[122]

NOTES

1. All works by Heidegger will be cited according to the abbreviations listed in the Bibliography of this book. In addition to *Being and Time* (hereafter SZ) the following works by Heidegger contain important contributions to the concept of time: *Prolegomena zur Geschichte des Zeitbegriffs* (PGZ), *Logik. Die Frage nach der Wahrheit* (LFW), *Die Grundprobleme der Phänomenologie* (GP), *Phänomenologische Interpretation von Kants Kritik der reinen Vernunft* (PhIK), *Metaphysische Anfangsgründe der Logik im Ausgang von Leibniz* (ML), *Einführung in das akademische Studium* (EAS), and *Die Grundbegriffe der Metaphysik* (GM). For Heidegger's view on the problem of time before 1927 cf. "Der Zeitbegriff in der Geschichtswissenschaft" (1916), in *Frühe Schriften* (FS), pp. 413ff. For a summary of the lecture "Der Begriff der Zeit" which was delivered on July 7, 1924 cf. Becker, *Mathematische Existenz* which first appeared in *Jahrbuch für Philosophie und phänomenologische Forschung*, 8(1927). For the latter see Theodore J. Kisiel, "Der Zeitbegriff beim früheren Heidegger (um 1925)," in *Phänomenologische Forschung*, 14(1983), pp. 192ff. and Thomas J. Sheehan, "The 'Original Form' of *Sein und Zeit*: Heidegger's 'Der Begriff der Zeit'," in *Journal of the British Society for Phenomenology*, 10(1979), pp. 78ff. As far as Heidegger's conception of time after the "turn" is concerned, the essay "Zeit und Sein" which appeared in *Zur Sache des Denkens* (SD), is still the most important text.

2. Cf. SZ, p. 426.

3. Cf. LFW, § 19.

4. Cf. SZ, pp. 422ff.

5. Cf. SZ, p. 426.

6. Cf. *Kant und das Problem der Metaphysik* (1973), p. *xiv*, LFW, p. 200, 400.

7. Cf. KPM, pp. 182-198; LFW, pp. 338ff.; PhIK, pp. 150ff., 269, 390ff. For a more careful presentation and justification of this thesis which cannot be brought into harmony with Kant's own views, cf. Marion Heinz, *Zeitlichkeit und Temporalität im Frühwerk Martin Heideggers*. Würzburg: Königshausen & Neumann, 1982, pp. 121ff.; K. Düsing, "Objektive und subjektive Zeit. Untersuchungen zu Kants Zeittheorie und zu ihrer modernen kritischen Rezeption," in *Kant-Studien*, 71(1980), pp. 1ff.; Carl-Friedrich Gethmann, *Verstehen und Auslegung. Das Methoden-*

problem in der Philosophie Martin Heideggers. Bonn: Bouvier, 1974.

8. Cf. SZ, p. 1.

9. SZ, p. 11.

10. KPM, p. 216.

11. Cf. SZ, p. 19.

12. Cf. SZ, p. 18.

13. Cf. SZ, p. 26.

14. Martin Heidegger, "Brief an Richardson," in William J. Richardson, *Heidegger: Through Phenomenology to Thought*. The Hague: Nijhoff, 1963, p. *xiii*.

15. Cf. SZ, p. 13.

16. Cf. SZ, p. 17.

17. Cf. SZ, pp. 37f.

18. Cf. SZ, § 7.

19. Cf. SZ, § 32.

20. Cf. Gethmann, *op. cit.*, pp. 259-260.

21. Cf. SZ, p. 114, 133.

22. The analyses of historicity and of the "everyday" understanding of time also serve the additional purpose of making a contribution to the eksistential determination of the self; the authentic self is that coming-to-pass of temporality which temporalizes itself. Inauthentically Dasein understands temporality only as the now-time that just passes away, whereas it misinterprets its own self as something that persists in time, thus as a substance.

23. Cf. SZ, p. 405.

24. Cf. SZ, p. 436.

25. For this issue cf. Heidegger's marginal notes *a* and *b* in the Klostermann edition of SZ of 1977, p. 41 and in WG, p. 41; cf. also the concluding remarks by von Herrmann concerning *Being and Time*, in the same edition of SZ, pp. 581-82 and *The Basic Problems of Phenomenology*, in GP, pp. 472-73.

26. Cf. SZ, p. 17, 333, 436.

27. Cf. SZ, p. 326.

28. Cf. SZ, §§ 60, 62.

29. Cf. SZ, p. 324, 327.

30. Cf. SZ, p. 325.

31. ML, p. 266.

32. Cf. SZ, pp. 284-85.

33. Cf. SZ, pp. 325-26.

34. Cf. SZ, p. 326.

35. Cf. SZ, p. 329.

36. Cf. SZ, p. 326.

37. SZ, p. 328-329.

38. Cf. KPM, p. 114.

39. GP, p. 378.

40. ML, p. 265.

41. Cf. SZ, pp. 390-91, 409; ML, p. 267.

42. SZ, p. 390.

43. GP, p. 428.

44. Cf. KPM, § 22, p. 192; GP, p. 379, 448.

45. SZ, p. 365.

46. Cf. SZ, p. 331.

47. Cf. SZ, p. 328.

48. Cf. SZ, pp. 328-29, 350; ML, p. 268.

49. For the concept of stretchedness cf. SZ, pp. 374-75; GP, 382.

50. Cf. ML, p. 266.

51. SZ, p. 351.

52. For this issue cf. SZ, p. 350: "...in every ekstasis, temporality temporalizes itself as a whole; and this means that in the ekstatic unity with which temporality has fully temporalized itself currently, is grounded the totality of the structural whole of eksistence, facticity, and falling--that is, the unity of the care-structure."

53. Cf. PGL, p. 422.

54. Cf. W. Müller-Lauter, Möglichkeit und Wirklichkeit bei Martin Heidegger. Berlin: de Gruyter, 1960, p. 54, 57 n. 3; D. Sinn, "Heideggers Spätphilosophie," in Philosophische Rundschau, 14(1967), pp. 81ff.

55. Cf. SZ, § 69b.

56. Cf. SZ, p. 158.

57. A. Rosales, Transzendenz und Differenz. Ein Beitrag zum Problem der ontologischen Differenz beim frühen Heidegger. The Hague: Nijhoff, 1970, p. 214.

58. Cf. SZ, pp. 341-42.

59. Cf. E. Tugendhat, Der Wahrheitsbegriff bei Husserl und Heidegger. Berlin: de Gruyter, 1967, p. 316; F.-W. von Herrmann, Die Selbstinterpretation Martin Heideggers. Meisenheim am Glan: Hain, 1964, pp. 155ff.; A. Rosales, Transzendenz und Differenz. p. 116.

60. SZ, p. 329.

61. ML., p. 267.

62. Cf. ML, p. 267.

63. Cf. SZ, p. 328.

64. For the inauthenticity of the present as such cf. SZ, p. 347: "The more inauthentically the present is--that is, the more making-present comes toward itself...."

65. GM, p. 224; cf. also SZ, p. 338; for the expression "Augenblick," cf. Otto Pöggeler, *Der Denkweg Martin Heideggers*. Pfullingen: Neske, 1963, pp. 209-10; D. Sinn, *Heideggers Spätphilosophie*, pp. 104ff.; F.-W. von Herrmann, "Zeitlichkeit des Daseins und Zeit des Seins. Grundsätzliches zu Heideggers Zeit-Analysen," in *Philosophische Perspektiven*, 4(1972), pp. 198ff.

66. Cf. Marion Heinz, *Zeitlichkeit und Temporalität*, pp. 135-36.

67. Cf. SZ, pp. 359-60.

68. Cf. SZ, p. 186, 348.

69. Cf. SZ, pp. 338ff.

70. GP, p. 398.

71. GP, p. 399.

72. GP, p. 323.

73. GP, p. 436; in the context of the analysis of Dasein Heidegger always uses the German terms *"zeitlich," "Zeitlichkeit,"* etc. for temporal and temporality, etc.; on the other hand, at the level of the interpretation of the meaning of Being as such he always uses the corresponding Latin terms *"temporal," "Temporalität,"* etc. (Cf. GP, p. 433). This deliberate use of these terms is not yet found in LFW; cf. for instance LFW, p. 199. The German term *"temporal"* is translated as "Temporal" and *"Temporalität"* as "Temporality," as suggested by Hofstadter in the translation of GP.

74. Cf. GP, pp. 81ff.

75. GP, p. 95.

76. Cf. GP, pp. 101-102.

77. Cf. GP, pp. 423ff.

78. Cf. GP, pp. 424-25.

79. Cf. GP, p. 250, 417.

80. Cf. GP, pp. 435ff., 439, 443-444.

81. ML, p. 272.

82. Cf. Marion Heinz, *Zeitlichkeit und Temporalität*, pp. 178ff.

83. Cf. GP, p. 453.

84. For what follows see GP, § 21b.

85. GP, p. 448 (italics added).

86. For this term see GP, p. 89.

87. Cf. GP, pp. 465-66.

88. This is also the reason why Heidegger can designate temporality altogether by the term "Temporality," insofar as it, taken as a whole, renders possible the understanding of Being. Taken in the narrow sense only the horizonal schemata constitute the Temporality of Being. Cf. GP, p. 389, 444, 459.

89. GP, p. 321.

90. Cf. GP, p. 322.

91. Cf. GP, pp. 445ff.

92. Cf. Marion Heinz, *Zeitlichkeit und Temporalität*, pp. 188ff.

93. Being taken as reality is thus ontologically founded on the Being of Dasein taken as care; cf. SZ, p. 211.

94. Heidegger declares that Kant's ontology insofar as it is an ontology of what is present-at-hand, really is a regional ontology of the temporal ontology proposed by Heidegger himself. Cf. PhIK, p. 200, 426.

95. Cf. KPM, p. 229; GP, pp. 26-27, 31.

96. Cf. SZ, p. 402.

97. With respect to this realm of themes also see GP, § 19; for a criticism of Heidegger's effort to present a derivation of the "everyday" conception of time, cf. Thomas M. Seebohm, *Zur Kritik der hermeneutischen Vernunft*. Bonn: Bouvier, 1972, pp. 58ff.

98. Cf. SZ, p. 329, 405.

99. Cf. SZ, p. 332.

100. Cf. SZ, p. 335, 370ff.

101. Cf. SZ, p. 404.

102. SZ, p. 405.

103. Cf. SZ, § 79.

104. Cf. SZ, § 80.

105. SZ, p. 421.

106. Cf. SZ, § 81.

107. SZ, p. 407.

108. SZ, p. 408.

109. Cf. SZ, p. 408.

110. SZ, p. 409.

111. Cf. SZ, p. 411.

112. Cf. SZ, pp. 411ff.

113. Cf. SZ, p. 414.

114. Cf. SZ, p. 420.

115. Cf. SZ, pp. 416-17.

116. SZ, p. 421.

117. Aristotle, *Physica*, Δ, 11, 219b, 1ff.

118. SZ, p. 421.
119. Cf. SZ, p. 421.
120. Cf. SZ, pp. 422ff.
121. Cf. SZ, p. 424.
122. Cf. SZ, p. 425.

11. Graeme Nicholson

EKSTATIC TEMPORALITY IN SEIN UND ZEIT

In giving talks to university circles during the early years of his career, Heidegger spoke more than once on the subject of time. There was, for instance, the lecture given in Freiburg in 1915, and published in 1916, "The Concept of Time in Historical Studies."[1] Then there was the paper given at Bultmann's famous theological colloquium, in July, 1924, after Heidegger had moved to Marburg in autumn 1923.[2] This text affords evidence that a close connection between Heidegger and the Bultmann group had already been forged less than a year after his arrival. About a year later again, Summer Semester 1925, we have the Marburg lecture course, *Prolegomena to a History of the Concept of Time*,[3] the course in which Heidegger gave the first major presentation of his own philosophical thought, and which foreshadows major themes of *Sein und Zeit (SZ)* (1927). My concern in this paper is really with *SZ*, but there is one observation I should like to make at the start, prompted by a look at this early work.

Beginning in the late 1920's, Heidegger was read in Germany and abroad mainly as a thinker preoccupied with the dilemmas of human existence, or what he called Dasein. It was a later generation, reading him in the 1950's and subsequently, that corrected the initial reception, seeing him as a thinker who had always been in search of the meaning or the truth of being, not just of our human existence. I think that both readings of Heidegger tended to assume that his interest in temporality or time was in a way preliminary, or subordinate. He studied temporality, but we thought that was only to throw light on some key features of Dasein's existence; he studied time as such, but we thought that was just to elucidate being as such. I do not mean that the priorities should be *reversed*! Rather I mean that the picture of a single ultimate topic, and of a penultimate goal or secondary topic, common to both readings, is not helpful in either case. If

208

we look at the 1915 lecture, we see that it reaches its goal in differentiating two articulations of time or temporality, one employed in natural science and one employed in historical research. (The second variant never exhibits the uniformity or homogenization found in the first variant with its commitment to the measurability of its phenomena.) But these differences of time itself are the key focus of the lecture. If we turn to the 1924 lecture, we would be equally able to say that an analysis of time is launched so as to throw light on Dasein, or put it the other way around: that we study Dasein in order to gain an insight into time. The very difference between these subjects is suspended: *"Das Dasein...ist die Zeit selbst,"* we are told at the climax of the short lecture. As for the 1925 lecture course, it is true that it barely gets to the topic of time annnounced in the title, but what is notable is that Heidegger rehearses major chapters of the analysis of Dasein under the heading "A preparatory description of the field in which the phenomenon of time becomes visible" (*G*, XX, 183). Of course, this is not really different from the plan of the published portion of *SZ*, and I mention it only as a means of rectifying the imbalance of decades of Heidegger-interpretation. We have re-worked the themes of Dasein and being exhaustively, but left the themes of temporality and time relatively untouched.[4] And yet Heidegger exerted the full force of his genius upon these topics in *SZ* (and, by the way, achieved an originality, a depth of treatment, that was not yet evident in the three earlier texts).

Dasein and Care. The topic of the temporality of human existence is introduced in *SZ:* § 65 as a further elucidation or prolongation of the view that our very being can be comprehended as care *(Sorge).*[5] It was the First Division, § 41, that proposed that view, and we need to take a preliminary glance at it. In context, Heidegger elucidates it by way of the experience of anxiety, but we need not explore the particulars of that; what we shall need is a sense of the general force of his conception of care, then of its three-fold composition or constitution.

Further back in the book, before the account of anxiety, there are many sections and studies pointing forward to the thesis of § 41. One key passage is § 12, in which Heidegger shows that Dasein's basic constitution is Being-in-the-world. In this context, the term "in" does not mean the spatial relationship of being contained, but rather a mode of dwelling that brings both a habituation to something and a concern or preoccupation with

something (p. 54). To regard such a condition as fundamental to
our constitution is to reject self-sufficiency and stable immanence
as aspects of the constitution of Dasein. Later, for instance in
§ 25, Heidegger indicates that his view also resists those views of
"selfhood" and "subjectivity" that lay stress on our self-reflexiv-
ity and identity. A few pages later, in § 26, he shows that in
our daily life we never cease to experience the kind of dealing
with things that he calls *Besorgen* or preoccupation, or the kind
of relationship to other Dasein that he calls *Fürsorge* or solicitude. The reference to these structures of experience prepare us
for the root idea of *Sorge* or care that is contained in them. Just
by virtue of the sort of beings that we are, we are constantly
exiting from self to be concerned with this or that, and yet
thrown back in all cases upon the resources that we possess. To
explain the semantics of the word "Being" in the phrase, "The
Being of Dasein is Care" is beyond my scope here; the indications
drawn from earlier pages must suffice for present purposes. We
must see now what elements are present in care.

   The first component, described at length and more than
once, has two principal names. The first name, *Verstehen*, signifies that thrust of every Dasein towards possibilities--of being
this or being that--which also discloses possibilities for all the
things that might be encountered in the world. The possibility of
myself to be this or that is not merely entertained or contem-
plated, but projected (*Entwerfen*, pp. 143-148) and so the word
Verstehen should be translated "projection" rather than "understanding." That projection, so defined, enters as one component
into care is made clear on pp. 182 and 192, as well as in § 65
and § 68 to which we turn later on. The power of projection in
Dasein is closely associated with the structure designated with a
second word, *Existenzialität* (see for instance, pp. 191-192 and
327-328) which I shall just transliterate into English here; whereas on pp. 12 and 13, for instance, this signifies the whole of the
constitution of Dasein's being, on later pages, such as the ones
cited immediately above, it forms one component in our being,
closely connected with projection. (But exactly what is the connection between them? Why does Heidegger use these two words?
We can see from pp. 143 and 191 that while projection is a power
of disclosure in us, existentiality is that which becomes disclosed
by projection. It is the aspect of Dasein's being that becomes disclosed through Dasein's projection.) The two may, however, be
treated as one in this essay, for we shall find in a moment a
name which can be used for both.

There is a second component of care which is also referred to by two major names (and a host of less prominent ones as well). Even as Dasein projects possibility for himself and for whatever he may encounter, he has already found himself in one locale, adopting some posture; he has already been exposed to the effects of an environment and registered them. To be affected by the world is not a contingent happenstance for Dasein, but a priori his condition as being in the world, so that affectedness springs in fact from an ontological structure prompting the affective response, a structure which would be called in English affectivity but which in German is called *Befindlichkeit*. (In early texts, including the 1924 Marburg lecture on time, Heidegger introduced his unidiomatic abstract noun as a translation of *affectio*, found in various texts of Augustine). Affectivity qualifies all of Dasein's projecting. Moreover, in various other texts we discern that moods (*Stimmung*), all attunement-to-things (*Gestimmtsein*) and the thrownness that has already located Dasein in a world prior to all choosing *(Geworfenheit)* are aspects of the structure of our affectivity--see pp. 134-139, 181-182, 191-192. But there is a second principal word used for this structure, particularly in § 39 and § 41 as well as in § 65 and § 68 later in the Second Division, which we can also just transliterate into English: *Faktizität*. As with the first pair of terms, we can say here that affectivity is one of Dasein's modes of disclosure, and that what becomes disclosed primarily through it is an ontological determination, a structure of Dasein's being, his facticity--see pp. 135, 139, 182.

There is a third component of care, referred to by the colorful term *Verfallen* that summons up images of the fall of man in Christian theology and Spenglerian themes of the collapse and decline of civilization. Even as Heidegger insistently denies these overtones, and never ceases directing the reader towards an ontological reworking of the term (see, for instance, pp. 175 f., 179 f.), he cannot cancel out the resonances the word actually has. If we examine what the word is intended to signify in § 38 and § 41, however, what emerges is the tendency of Dasein to become lost in the multifarious business of an everyday world, to be diverted from all urgency of decision, to get by with ease and conformity, in short, the perpetual tendency to become compromised in the banal, the trivial, the materialistic or the orthodox. Where this loss is inscribed in Dasein himself as his normal fate, let us imagine it as Dasein's constant compromise, and let us try to use that word to translate Heidegger's *Verfallen*. (While projection and affectivity do constitute a triad of which discourse or talk--*die Rede*--is the third member, there is no overlap between *Rede* and *Verfallen*.)

Now there is little doubt that our triad adds up to an intelligible sketch for a philosophical anthropology. While constantly projecting possibilities for himself--and thus being always *out in advance of* his current state--Dasein is nevertheless pervaded by innumerable influences and responsive to them--which is to say he is *already encompassed in* an environment or world even as he projects a possibility. In framing his projects, Dasein has the perpetual tendency *to follow the lead of* things, persons and institutions to which it is easy to submit; from the start, then, compromised with that which presents itself to him most readily and immediately. Being out in advance of self--yet encompassed already in (a world)--so as to be caught up with (the available and the immediate):[6] this is a single structure, despite its three components, the unity of which, as an a priori expression of the intentionality of Dasein, becomes more visible to us when we suppress the object--terms that are in the parentheses. And where we saw two terms for the first element in the structure of care, we shall now use the new term "in-advance-of-self" to stand for either of them, and, in place of the two earlier words for the second element in the structure, we shall use the term "being-already-encompassed-in-." For the third element we shall also use the new term "being-caught-up-with-." Heidegger uses hyphens this way to keep our attention fixed on the incompleteness of any one of the elements of the structure, taken by itself.

Care and Temporality. Before his introduction of temporality in § 65, Heidegger offers a brief resume of the consequence his notion of care has had for the notion of selfhood in § 64. Because Dasein is in every case in advance of self, and encompassed in (the world) and caught up with (what is immediately available), it follows that traditional theories of selfhood that are based on substance are in error. Moreover, he holds that Cartesian and Kantian theories of the ego as a subject have not escaped the influence of ancient tradition. They have not grasped how, by virtue of its care-structure, the human being is inevitably outside of itself, escaping the stable immanence of coinciding with self.

With the essence of the care-structure thus established, Heidegger turns, in what is the pivotal turn of the entire book, to the question whether it is possible to interpret the structure of care any further. (Obviously, much has been done in the sections separating § 41 from § 65, but as we shall see presently there is one key manoeuvre of interpretation that can be accomplished only now.) Before we probe the deep philosophical issue posed by § 65, let us look at the rudiments.

It is Heidegger's argument that the three elements composing care need to be traced further back to that which makes them possible, or gives them a grounding or foundation. Therefore, he asks first how it is possible that there should be projection. How is it that a possibility should be projected or anticipated? And in particular how is it that a possibility for myself should be pro-jected by myself? His answer is that my being towards the possi-ble is a bearing that only becomes possible for me because I can permit the approaching of a possibility. That which arrives or comes towards us *(Zukommendes)* can do so because a zone of approach or advent is held open for it; such a zone of approach, then, is also what makes it possible for us to permit the approach of our possibility, to project. Projection, or being in advance of self, is made possible by the *Zukünftigkeit*--capacity to make approach--of the *Zukommendes*--that which approaches--and that *Zukünftigkeit* or the power of making approach is the original or primordial phenomenon of the *Zukunft*, the future (p. 325).

It is not that our projection constitutes the future. Rather the power of a possibility to approach us makes our projection possible, but this power is the primordial future. Let us look at the next paragraph of text (pp. 325 f.). Just by virtue of per-mitting a possibility for itself to approach it, Dasein will not evade or escape the fact of being now what it has hitherto become. It takes it upon itself to be whatever it is that life has made it be. To assume this persona--and that may be a burden for it--so far from being ruled out through an opening towards a coming possibility, is actually implied by the latter. Where my possibility is approaching me, I assume the burden of what I have been or been made to be: I am my having-been. This is how a Having-been *(Gewesensein* or *Gewesenheit)* springs, as Heidegger says (p. 326), from the future. It is on p. 327, not here in the paragraph I've been summarizing, that Heidegger points out that the Having-been he has deduced constitutes the ground or foun-dation for that being-already-encompassed-in-(the world) that was the common character of our affectivity and our facticity. The second element of care is founded in a Having-been.

In the next two paragraphs, Heidegger points out that any Dasein that permits its possibility to approach it, and that takes upon its shoulders its Having-been, will of course encounter a

manifold of available things in the immediate surroundings. They
are immediately present *(Anwesenden)*, but only because of
Dasein's act of *Gegenwärtigen*, encompassing them within the
circle of its own concerns or making them present. Yet Dasein's
making (things) present springs only out of the futurity we have
described as the permission of possibility's approach and out of
the past we have described as the assumption of a past, a
Having-been. While presence is deeply connected with the things
that appear immediately around one, it is not in its roots an
attribute or achievement of the things, but an act of Dasein,
making--(things)--present. On p. 327, Heidegger says that the
third element of Dasein's care, our being compromised with
(things) or caught up with (them) is founded in the structure we
have now surveyed, making-present.

Insofar as there is permitted the zone of the approaching in
which a possibility may confront that which, because I have been
it, I am, and thereby permits the making-present of things, there
is constituted a single phenomenon, of unitary structure, which
Heidegger calls temporality. His formula for it includes a noun for
the future, signifying the primacy of the future in the structure,
with modifiers, linked by a hyphen, signifying past and present:
die gewesend-gegenwärtigende Zukunft. It is the temporality of
Dasein we have in mind here, and so we are free to translate
Heidegger's phrase with the terms we have already introduced:
the Having-been and Making-present Approaching. (Later we'll
see modulations of this formula.) In this is founded the ground of
the possibility of Dasein's care. And now we can identify a key
question to pose to Heidegger's thesis. Is it human care which
accounts for the characteristic features of human temporality? Or
is it, as Heidegger says, human temporality which accounts for
the characteristic features of human care, serves as their foundation?

It might be thought, since Heidegger is basing his theory
upon the view that human *being* is constituted by projection,
affectivity, and so on, and since these ontological determinations
show up in § 65 reinterpreted in temporal terms, that he is in
effect basing Dasein's temporality upon Dasein's being. After all,
minerals and insects do not occupy time the same way that human
beings do. Is that not because they have a different mode of
being? Can we not say that there are distinct sectors of entity,
and that in each sector temporality appears differently, owing to
the mode of being found in each? The treatment of temporality
would be made relative to the sector in question. And there is

another point. It has often been the approach of phenomenology in this century to underline what is specifically human about the space that opens out to us, of the time that opens out to us, or the way in general that phenomena of nature take on attributes in the light of our experience. In this case, it would be the human experience or consciousness of time that is treated.

It is quite plain from what we have already summarized that what we are studying is the kind of temporality that is characteristic of human experience. We have seen how possibility approaches us insofar as we are projecting ourselves; we have seen how it is given to us to assume the burden of our Having-been; and so on. Nothing in these descriptions is applicable outside the range of human experience, certainly not to the rotation of planets, the duration of radio-active materials, or the life cycle of insects. Our theory has been relativized from the start to the human scene and to the world of which we are conscious. On that assumption, then, could we not immediately identify the "temporality" of Dasein with the "time-consciousness" of Dasein? Might we not infer that such features of our temporality as the approach of possibility and the weight of Having-been are features of "human time" in the specific sense that they are the attributes time itself would appear to have to a being like ourselves preoccupied with care? This is a further way in whch we could stress the difference between the theory of the temporality of Dasein and other theories referring to other entities. Specifically different to Dasein is the actual consciousness of time. Radio-active minerals show no tendency to assume the burden of their own past, for instance. Insects exhibit no projection of possibilities. Specific features of our *consciousness* would account for the many specific features of the temporality we experience as human beings.

The temporality of our existence need not be limited to our consciousness of heartbeats, clocks, and the like, nor to our sense of boredom or anxiety about the future. Temporality could, in other words, be relativized to categories of entity (mineral, insect, Dasein), quite in a general way, by "sectors" of entity. Then, in addition to that, we would find a time-consciousness in Dasein, affected by human wishes, hopes and cares. The common point between both these approaches is that specific attributes of the *temporality* of Dasein arise from the specific mode of *being* of Dasein, care. They would, then, be in their roots not aspects of temporality itself, but reflections of Dasein's care.

And yet, plausible as it might seem in view of some readings of Heidegger, his approach cannot be interpreted as a sector theory of that sort any more than as a phenomenology of temporal experience. For one thing, the sector theory is still moving in the same framework as Aristotle's *Physics*, Book IV, with its thesis that time is an abstract aspect of the movement of certain phenomena. Merely to differentiate classes or sectors of phenomena, and then derive corresponding temporal differences, does not move away, as Heidegger intends, from the fundamental principles of Aristotelianism. Moreover, if it is proposed to elucidate human temporality through characteristics of human being, the problem of method will stand out in high relief: from what leads shall we derive the theory of human being? And is it not evident that Heidegger wishes to cast light upon human being by way of the study of human temporality?

It is to confuse the order of explanation to view our temporality as deriving its characteristics from our being, care. Instead it is necessary to view human care as the expression of human temporality. And in particular the three-fold constitution of care stems from the three-fold constitution of temporality.

Temporal Predicates and Their Unity. The title and the contents of § 65 both confirm that, in the relationship between care and temporality, it is the latter that is the foundation of the former. The title, "Temporality as the Ontological Meaning of Care" is given a careful interpretation on pp. 323–325, in which Heidegger makes it clear that when X is the meaning *(Sinn)* of Y, then (a) we can understand Y with reference to X (where "understood" has the technical sense of "being projected or designed" *(entworfen)*, (b) that the unity or wholeness of Y is brought about by X, and (c) that it is X that makes Y possible. Thus, we are to find that care is intelligible to us by way of temporality, that care is a unity or totality because of temporality, and that it is temporality that makes care possible.

Still, if the matter were quite as simple as that, we would find it hard to justify Heidegger's treating care as the being of Dasein in § 41, and only turning to temporality as the meaning of the being later on, in § 65. It is fairly clear from what we have already seen that the study of care has guided us in our approach to temporality, so that, as far as point (a) at least is concerned, the relationship between the two is reciprocal. A scrutiny of the text of § 65 and especially § 68 prompts the following resolution of the matter. There are three parts of temporality, known as ekstases (more on that term below), and they stand in a very special form of unity illustrated by the formula quoted above in which a noun for the future was qualified by terms signifying past and present joined by a hyphen. But there is a series of substitutions which can be made for the three terms. We discover four or five different modulations of futurity, pastness and presence, four or five different formulas for temporality, therefore, with members linked in the particular way we have seen. *The particular substitutions made, new names for each part of time, are derived from the analysis of care. But the overarching form of ekstatic unity into which the substitutions are brought does not stem from the analysis of care, but is imposed upon care by the analysis of temporality.* The unity, therefore, springs from temporality, but the members brought into unity are derivative of care. Let me now seek to prove that way of reading the text.

The temporality of Dasein must be treated phenomenologically, which is to say according to the way it generally and initially appears, in the everydayness of Dasein's life. So the chapter that runs from § 67 to § 71 treats temporality and everydayness, and § 68 in particular gives a rich profusion of modes in which temporality shows up in everyday experience. These modes are all to be qualified as either "authentic" or "inauthentic" forms of future, or past, or present, because in everyday existence Dasein never escapes the lure of an inauthentic life, the lure of compromise or *Verfallen*. Existing not in view of one's own possibilities but according to well-trodden paths, foreshortened possibilities, and preoccupation with the immediate and available, will bring with it in every case an inauthentic bearing to the future, the past and the present. And we can be more specific on this score. Our existence can be divided into a great many distinct structures, and it is possible to show with each of these structures how an inauthentic bearing shows up in a future-orientation, a past-orientation, and an orientation in the present. § 68 focusses upon the constituent elements of Dasein's disclosedness-- projection, affectivity, compromise and talk--showing how with each one of these structures an authentic bearing towards future, past and present can be discriminated from an inauthentic bearing. One of the problems of this text, which would need a longer treatment than I can give it here, is that things are folded in upon themselves in a curious way here. The text treats the temporality of projection, of affectivity, of compromise and of talk, but without observing that the first three of these have already entered as elements into care, and that care itself reappeared,

reinterpreted as temporality with its three parts. Another difficulty is that *Verfallen*, or compromise, is the root of Dasein's
inauthentic existence, and therefore implicit in every structure
that takes an inauthentic form. Why then is there a special treatment of the temporality of compromise alongside the temporality of
projection, of affectivity and of talk, domains where inauthenticity
appears? Nevertheless, despite these questions, we do find striking and concrete descriptions here, and I shall give a brief
catalogue of Heidegger's innovations.

Projection, or existentiality, reinterpreted as being out-in-
advance-of-self, proved to be the prototype of Dasein's futurity,
we saw in § 65; or better, was made possible by the primordial
form of futurity, the power of possibility to approach us. The
temporality of projection will therefore be characterized most of all
by a bearing towards this possibility, or this future. And where
it is my own possibility in the proper sense that I permit to
approach--i.e., where I exist as myself or authentically--the
futurity of my projection will be *Vorlaufen* (p. 336), a Running--
ahead--towards--, or *Preparing*, the Approaching. But a projection that does not frame self-generated possibilities, but simply
looks for what the world will offer, is inauthentic in its bearing
towards possibility, and is called *Gewärtigen* (p. 337), or *Expectation*. The future towards which Dasein looks is the *Expected*, or
(we can say in English with the plural) his *Expectations*. One
form of *Gewärtigen* is Waiting or Awaiting *(Erwarten*, p. 337).

It is projection we are discussing now, so its temporality will
be characterized most of all by its ways of bearing towards the
future, but not solely by that, since a form of the present and a
form of the past will be found both for an authentic projection
and an inauthentic one. An authentic projection that prepares a
possibility will experience decision, and will be open to discerning
crisis situations, ready therefore to experience every present time
as *Augenblick* (p. 338), or a *Juncture*. But where the possible is
merely expected, the present time is merged or blended into the
Fortuna that is approaching, and present time is good for no
more than an occasion for the Expected to materialize, a mere
Gegenwärtigen (p. 338), a Making-present, or *Realizing*, of the
Expected. Finally, where Dasein can seize the Juncture to Prepare the Approaching, he will not evade being his Having-been,
taking up again what the past has done to him, so an authentic
projection will live the past as *Wiederholung* (p. 339), or *Repetition*, an act that relives the past under the category of possibility, a past which may be unique to oneself, or on the other hand

a past lived through by another, but capable of being a possibility for my Repetition. But where Dasein lives by Expectation, in a present only as an occasion for Realizing Fortuna, its past will be ignored and cast off: *Vergessenheit* (p. 339) or *Oblivion* is the past of inauthentic projection, and as a variant of that, the deficiency of Oblivion, there is the Recollected, Remembered past, *die Erinnerung* (p. 339). The six principle modes (Expectation and Preparation, Juncture and Realizing, Repetition and Oblivion), together with the two subordinated modes (Waiting and Recollection) are predicable of various parts of time, and I shall treat them so and speak of them as time-predicates. There are a few further individual time-predicates listed in later sections, and we shall refer to a few of them, but the ones we have catalogued here are the most important ones for *SZ*. Now we must turn to the question how these predicates are woven together into unities.

The Ekstasis of Temporality. In the section above on Care and Temporality I pointed out, not only that each component of care was being traced back to one of the parts of temporality, but also that the latter three had definite connections among themselves. It was because a future possibility was able to make an approach that Dasein took up the task of being its Having-been, and it was on account of both these relationships that Dasein would undertake the Making-present of entities. We already cited the formula for temporality, *die gewesend--gegenwärtigende Zukunft*. This formula makes it clear to the reader that the three components cohere into a unity, but also clear that the unity in question is a synthetic unity, the convergence of three members, and is not the unity of a single line, measure or dimension. On pp. 326-328 Heidegger seeks repeatedly to differentiate the structure he is talking about from our common representations of time, and in the paragraph spanning pp. 328 and 329 he draws an explicit contrast between the unity that arises when an Approaching prompts and induces a Having-been such that both permit a Making-present from the unity found in the linearity of a mere sequence, such as a sequence of Now's. In fact he claims here that such a sequence, which constitutes the main feature of the common representation of time, has arisen through the levelling-out of the key contours of the *gewesend--gegenwärtigende Zukunft*. But why is it, in fact, that each of these members enters into a synthetic unity?
    Futurity is essentially the Approaching-towards...(*Aufsich-*

zu); Having-been is essentially the Returning-upon...(*Zurück auf*); Making-present is essentially Granting-presence-to... (*Begegnenlassen von*) (p. 328). Each of the elements is characterized by what we shall call an Opening-out-to-(something). Heidegger tells us that this reveals that temporality is utterly and completely ekstatic (*das ekstatikon schlechthin*, p. 329). If ekstasis means being placed out of self, displacement, the absolutely and completely ekstatic is that which never possessed an immanence, never had a position other than displacement, or as Heidegger expounds it here, that which is "by origin 'outside-of itself' by its own very nature" (...*das ursprüngliche 'Aussersich' an und für sich selbst*, p. 329). Although Heidegger does not say so explicitly here, we can infer from such words as *schlechthin* and *ursprünglich* that whatever is touched by temporality, or participates in it, will acquire some degree of ekstasis from it. Futurity and the other two parts are utterly and completely ekstatic, and do not have that as a mere attribute, and so they are actually named, from here on, the ekstases of temporality. Each of them has as such, *an und für sich*, the character of being outposted from self. For that reason, each must perpetually enter into each of the others. Synthetic unity arises as a result of the outposting occurring in each of the three ekstases. Temporality is constituted as a unity by outflow from the three rather as the lighting of a stage may be composed of interplaying beams of different coloration from three sources. The quality of the synthetic lighting (*Lichtung*) can be resolved into three ekstases. But it is a reaching, an openness, a need, in the original ekstasis of each member, an endowment in each, that yields their exiting or outposting. It is not possible to trace the tendency of each to reach out of self back to some external prompting, as if, for instance, the tendency of futurity to permit the approaching possibility were the work of some particular possibility, as if an incoming possibility were the cause of the opening up to possibility. Opening for possibility must be foundational for any possibility. The same holds of the past and present ekstases: the outposting and opening have not been prompted by something which was in the past or something else which now exists in the present. Is it possible further to explain the self-exiting, the outposting, of the futural ekstasis or of the other two? Or must we accept ekstasis as an ultimate given? Why does time reach? To explain this is not possible within the compass of the Second Division of *SZ* "Dasein and Temporality," from which we are quoting now; the task of this division is confined to the impact upon

Dasein's being of his temporality. To trace the ekstasis of Dasein's temporality back to a more ultimate source in time itself was to be undertaken in the Third Division, "Time and Being," which Heidegger did not publish. (The 1962 lecture "Time and Being" does, however, make some contribution to answering the question: it speaks[7] of the "reaching" of presence--*Reichen*--out of all three temporal ekstases, and seeks to probe into this reaching, showing a series of aspects: lasting *(währen)*, granting *(geben)*, arriving *(erreichen)*, and so on.) In 1927, we know from later testimony,[8] this line of question forced Heidegger into an all-too-metaphysical mode of thinking, a sort of probing into founding relations among first and second "principles." Perhaps one could compare to Heidegger the speculations of F.W.J. Schelling on the *ekstasis* of reason in positing an existing world,[9] or the Plotinian overflowing of a sequence of hypostases from the One. Explanations of this style were of no use to Heidegger at this time, and prompted him instead to inquire again and again into the "essence of grounding,"[10] or "what is metaphysics?",[11] into the "end of metaphysics"[12] and the "principle of sufficient reason."[13] As far as the present text from *SZ* is concerned, even though the being of Dasein is made possible by the temporality of Dasein, still, that temporality lies much nearer to hand in a phenomenological study than does time itself. Time itself may well account for the ekstases of temporality; we have not shown that, in any case; what we can say is that human care does not account for the ekstasis of the future or of the other two divisions; rather, since they are primordially ekstatic, they transmit the quality of ekstasis to human care. The unity of interplay among the three ekstases is what binds up our being-in-advance-of-self with our being-encompassed-by-(the world) and our being-caught-up-with-(things).

Variations in the Temporality of Dasein. When appearing in the particular modulations of Dasein's everydayness, temporality becomes modified by Dasein's orientation. As I have been saying, the particulars of Dasein's posture in the world do not have the power to create or to alter the overarching unity and structure of temporality. Everyday modes of care vary in their intentions and goals, but no intention or goal can alter the circumstance that possibility moves in upon us and thereby gives us a past and a present. The kind of future, past, and present, however, is negotiable. Let us review briefly the variations found in § 68. A number of them are stated in the very pattern established in

§ 65, with three terms, one for each ekstasis, linked in the familiar way. There are other phenomena treated here, however, which are not reduced to the formula. I shall argue that it is easy to bring them into the formula.

It is an everyday existence we treat, and we begin with an everyday mode of *projection*, in which Dasein has not escaped the habit of its inauthenticity and compromise. Projection is a bearing most of all to the future, and so its temporality even in the everyday mode will be qualified most of all by the manner of bearing to the future. It will Expect a possibility rather than Preparing it, and exactly that will prompt a Forgetting of the past rather than Repetition, and an attitude of Realizing in the present rather than a decision at a Juncture. An ekstatic unity is thereby constituted, the Expectation that Forgets and Realizes, *das vergessend-gegenwärtigende Gewärtigen* (p. 339). We are not given such a schematic formula for the temporality of an authentic projection, presumably because the thrust of § 68 is towards inauthenticity. Nevertheless the materials have been provided by the text: an authentic projection would be temporally expressed (or temporalized) as a Preparation that Repeats at a Juncture, *das wiederholend-augenblickliche Vorlaufen*; that is how a conscientious pressing towards a possibility, that is ready to endure anxiety, would express itself in time.

In treating the second mode of our disclosing, or the second component of care, affectivity, Heidegger will naturally put the emphasis on the past. An inauthentic affectivity would be a kind of Forgetting (in the noun position) modified by the other two ekstases. He chooses to discuss this, however, with one instance of inauthentic affectivity, namely fear. It is (oddly!) fundamentally a Forgetting, but qualified as Expectant and Realizing, *gewärtigend-gegenwärtigendes Vergessen* (p. 342). Perhaps this would have been more suitable as a general description of the temporality of inauthentic affectivity, and then supplemented with concrete accounts of fear and other cases. Again, he does not give a schema for authentic affectivity, but he does offer a study of one case of it, anxiety, that discloses the repeatability of the past *(Wiederholbarkeit,* p. 343). It is not easy to decide whether final emphasis ought to fall on the ekstasis of the past or that of the future in the case of anxiety, for Heidegger says (p. 344) that it springs out of the future. If we stress the past ekstasis Repetition, that will at least signify how the past is apprehended as a possibility in this case, and if we qualify its mode of presence not as a Juncture (which is unsuitable for any affectiv-

ity, I think, even an authentic one) but as *gehalten* (p. 344)--
i.e., Steady--we could call it a Steadfast Preparatory Repetition,
die gehalten-vorlaufende Wiederholung. Indeed that might serve
not for anxiety in particular but for all authentic affectivity.

The temporality of compromise *(Verfallen)* will have, of
course, only the one branch, the inauthentic, for in Heidegger
there is no authentic *Verfallen*. But again we are given one
instance, curiosity *(Neugier)* which may illustrate the whole
structure. The emphasis falls now on the ekstasis of the present,
but curiosity "does not seek to bring its objects into presence in
order to dwell with them and understand them, but seeks to see
them only in order to see and to have seen. A way of realizing
that has become entangled in itself this way will be ekstatically
united with a corresponding future and past" (p. 346). The
future is a quasi-presence, the most impatient sort of Expectation. And the past is the most intense form of Oblivion--it is the
out-of-date, the old. We could propose a formula of ekstasis for
this instance of compromise, I think, *ein entlaufend-ungehaltenes
Gegenwärtigen*, an Evasive and Unsteady Realizing.

The last section of § 68, on the temporality of discourse,
does not provide us with sufficient detail to formulate a schema of
ekstasis. We do find one further variation in § 69, which plays a
major role in the rest of the book. The temporality of our daily
intercourse with the things around us *(das umsichtige Besorgen)*
proves to be a slight modification of the temporality of our
inauthentic projection. While the futural ekstasis prevailed in the
schema for projection, it changes places here with the ekstasis of
presence. And in place of a Forgetting we now have Retaining as
a mode of our relation to the past. In place of a *vergessend-
gegenwärtigendes Gewärtigen*, we have *gewärtigend-behaltendes
Gegenwärtigen* (pp. 353-355), a Realizing (of things) that is
Expectant and also Retentive. This normal, even banal, mode of
temporalizing is also what holds through when Dasein develops
into a scientist: we see, pp. 356-364, that even with all the
changes that are necessary to transform daily pragmatic encounter
with things into a theoretical way of studying them, this brings
no fundamental revolution in Dasein's mode of temporal existence.
(See also pp. 406-411).

Conclusion: Subjectivity and History. The variations we have
surveyed stem from the choices Dasein makes regarding the way
he will live in the world, and his dealings with things and persons. But what is not up to him to choose is that every form of

care, thrust towards one ekstasis in particular, will implicate him in the other two. The ekstasis of temporality lies well prior to any choice. The choices can affect one ekstasis, but the impact of one ekstasis upon the other two, and upon the interplay of the three, cannot be foreseen and calculated by Dasein. It is as if there could be different colorations of the three lights that converge to form a single synthetic *Lichtung* on a stage, but no choice of the circumstance that the three of them will come together. So our study of care introduces us to a matter whose scope outreaches care: the ekstases of temporality itself. The thinker can now begin to probe the very nature of ekstasis, why it is that each of the three is informed by the need to go out from self, why it is that the futural ekstasis has the primacy, why it is that that which is still stored up in the darkness of the future is a source for the past and the present. But these are questions which *SZ* did not succeed in treating.

What *SZ* was able to make clear is that the reign of temporal ekstasis over the choices we make accords with the place we occupy as finite beings in the world. Because the temporal ekstases outreach every Dasein, we all find ourselves located in the vast transmission of history. Our historicity is but a fuller expression of our temporality, and the reaching out of each of the historical ekstases determines the limits of the choices we have to make in our own generation. The place of individual human beings in history does not appear the same in Heidegger as it does, for instance, in the dialectical philosophy of Hegel and his successors. The power of the negative in a dialectic is the force that outstrips every positive position taken by a person, a group, or a civilization. But where temporality appears not as the driving power of the negative but as a reaching interplay constituted by a need, history will appear in the shape of possibility instead of necessity, and human choice, though it cannot cancel out temporality itself, can exercise concrete powers over future, past and present. These are some of the reasons that led Heidegger to undertake a critical engagement with Hegel in the penultimate section of the book, § 82. Our conscious aims may focus on some object in the past or the future or the present. The ekstasis of time does not constrain us in this respect. But it cannot permit our projection to outrun our affectivity, or our compromise to be quite without projection. The reach of time's ekstases bestows a unity upon our care. It also constitutes the bond that links generation with generation.

NOTES

1. "Der Zeitbegriff in der Geschichtswissenschaft;" the text as published has now been made available in Martin Heidegger, *Frühe Schriften* (Frankfurt: Vittorio Klostermann, 1972), pp. 355-375. See also the bibliographical note in the same volume, p. 376. A translation by H. S. Taylor and H. W. Uffelmann appears in the *Journal of the British Society for Phenomenology* 9(1978), pp. 3-10.

2. "Der Begriff der Zeit." Publication of this text within the collected works of Heidegger *(Gesamtausgabe; G)* is not foreseen for some time. A summary is available in Thomas Sheehan "The 'Original Form' of *Sein und Zeit:* Heidegger's *Der Begriff der Zeit* (1924)," *Journal of the British Society for Phenomenology* 10 (1980), pp. 78-83.

3. *G*, vol. XX, ed. Petra Jaeger (Frankfurt: Vittorio Klostermann, 1979).

4. Am I being unjust to the scholarship of the past three or four decades? Of the major studies of Heidegger in English, for instance, how many deal principally and explicitly with time? There is the important work by Charles Sherover, *Heidegger, Kant and Time* (Bloomington: Indiana University Press, 1971), but is there any in German, or any other at all? One could consult the main well-known and excellent anthologies of articles on Heidegger, e.g., those edited by Frederick Elliston (*Heidegger's Existential Analytic*; the Hague: Mouton, 1978), Michael Murray (*Heidegger and Modern Philosophy*; New Haven: Yale University Press, 1978), Otto Pöggeler (*Heidegger: Perspektiven zur Deutung seines Werkes*; Köln: Kiepenheuer & Witsch, 1969), and find very little in any of them explicitly devoted to time and temporality. [For additional works and articles on Heidegger's conception of time see the bibliography at the end of this book. (Ed.)]

5. I shall refer to the numbered paragraphs or sections of SZ with numbers and a paragraph sign, such as § 65.

6. This is how I translate the formula for care that appears on p. 192: *Sich-vorweg-schon-sein-in-(der-Welt) als Sein-bei (innerweltlich begegnendem Seienden)*.

7. *Zur Sache des Denkens* (Tübingen: Niemeyer, 1969), pp. 10-14. The English translation is available in *On Time and Being*, tr. Joan Stambaugh (New York: Harper & Row, 1972), pp. 9ff.

8. See particularly "Brief über den 'Humanismus'," now in *Wegmarken* (Frankfurt: Vittorio Klostermann, 1967). Translation

by F.A. Capuzzi in Heidegger, *Basic Writings*, ed. D.F. Krell (New York: Harper & Row, 1977).

9. See, for instance, the Introduction to Schelling's *Philosophie der Offenbarung, Sämmtliche Werke*, II, 3, 163 (Stuttgart: Cotta, 1856-1861). I am grateful to John Burbidge for references to Schelling's treatment of *ekstasis*.

10. "Vom Wesen des Grundes," appearing almost immediately after *SZ*, probed the conditions requisite for any positing of grounds, reasons or foundations. *The Essence of Reasons*, tr. T. Malick (Evanston: Northwestern, 1969).

11. The title of the famous Freiburg inaugural lecture of 1929. Translation, available in *Basic Writings*, by D.F. Krell.

12. A text called "Ueberwindung der Metaphysik," that appears in *Vorträge und Aufsätze* (Pfullingen: Neske, 1954) actually dates from the years 1936-1946, and may be read as a kind of programmatic statement of Heidegger's effort during that time to understand the end of metaphysics. A number of associated texts appear in *The End of Philosophy*, tr. J. Stambaugh (New York: Harper & Row, 1973).

13. *Der Satz vom Grund* (Pfullingen, Neske, 1957).

12. OTTO POGGELER

METAPHYSICS AND TOPOLOGY OF BEING
*IN HEIDEGGER**

Concerning a controversial matter such as M. Heidegger's thought we cannot go wrong by drawing upon new titles for its interpretation. Yet, the more titles are used the more misunderstanding there is, even though each title may shed certain light on the matter. As to the title of my essay "Metaphysics and Topology of Being in Heidegger," no one would deny that although he turns away from traditional metaphysics, metaphysics has been the issue of his thought. But what justification is there for conceding that he discusses the *topos* of Being? Indeed what does this expression mean? Is it a distinction of Heidegger's thought or primarily an indication of certain facts internal to it? Furthermore, how is, in Heidegger, the discussion of the region *(topos)* assigned to Being related to metaphysics?

If words such as metaphysics and "region assigned to Being" are not to be attached to his thinking merely as labels then we must begin by inquiring how does this thinking understand itself. This way perhaps it can become clear what the term metaphysics is supposed to mean, and also what Heidegger's discussion of *topos* of Being is all about. Indeed if these terms are not only

---

*This translation presents the text of a lecture Professor Pöggeler held before the members of the *Görresgesellschaft* in 1961 and subsequently published in *Philosophisches Jahrbuch der Görresgesellschaft* vol. 70/1, 1962, pp. 118–138. In the original the title of the essay is "Metaphysik und Seinstopik bei Heidegger." Throughout the original, Pöggeler mentions his sources parenthetically or does not mention them at all; we have followed the same practice. The English version was published first in *Man and World*, 8 (1975), pp. 3–27. Reprinted by permission of Martinus Nijhoff, Publishers B.V.

externally attached to Heidegger's thought then their relationship should become manifest from this thinking itself.

Heidegger's thinking begins with a question which has remained in a unique and exclusive fashion the guiding-question of occidental metaphysics. It is the question formulated by Aristotle at the end of classical Greek philosophy as : *ti to on?* What is a being in its Being? The "first" science in philosophy is charged with the task of grasping an entity as an entity, thus determining its Being. The Aristotelian thesis that "being is stated in many ways" holds true for this primary science.

For instance, I can say *what* this entity in front of me is (a speaker's desk) and *that* it is (it exists, is presently here) both of which may be expressed in some other way. In its Being an entity is comprised of its quiddity (what it is) as well as its existence (that it is). Furthermore, a being can be grasped, as it was in Scholastic philosophy, as *res* (thing), *unum* (one), *aliquid* (something), *bonum* (good, worth aspiring after), *verum* (true, i.e., manifest in its Being). *Res, unum, aliquid, bonum, verum* are ultimate and highest determinations which like *ens* itself transcend other determinations and are therefore convertible to *ens*. These highest determinations are called transcendentals.

Heidegger did not receive the doctrine of manifold ways of stating entity as what one may sometimes come across in the history of thought. Rather in the course of his own philosophical research he encountered the manifold meanings of Being as a problem. In his Dissertation *Doctrine of Judgment in Psychologism*, Heidegger inquired into the mode of Being and actuality of the logical phenomenon, and into the actuality of an objectively valid meaning which allows an object to become known and thereby a "true" object. He inquired here also into *ens tanquam verum* (a being as a true one). Pursuing neo-Kantian lines of thought and (above all else) Husserl's *Logical Investigations*, he sharply differentiated the mode of Being of the logical phenomenon--the *verum* from the mode of Being proper to the psychic phenomenon. Logical and psychic phenomena each have their own mode of Being. By attempting to conceive judgment, its forms and the logical phenomenon as originating in the psyche, psychologism basically misapprehends the mode of Being of the logical phenomenon. To psychologism, the mode of Being of the logical phenomenon as an independent mode is altogether unknown. Can this independent and irreducible mode of Being be proved for psychologism?

Heidegger maintains that nothing can be proved but certain things can be exhibited. In the area of transcendentals nothing can be proved insofar as these, implied in their name, lie above and beyond all determinations. They have no genus over them from which they can be deduced by applying the specific difference. In his dissertation (p. 95) Heidegger states:

> Here we are confronted perhaps with what is ultimate and irreducible, what excludes further clarification and brings any further inquiry to a stand-still.

It is the task of metaphysics to exhibit the ultimate and mutually irreducible ontological characteristics.

For Heidegger, however, these transmitted metaphysical ideas were not crucial; what is crucial is the question which they made audible to him. If being can be stated in many ways, should we not inquire into the unity out of which the multiplicity of the meaning of Being is articulated? Can we evade such an inquiry by maintaining that "Being" is the most general and undefinable concept? Is it necessary for the answer to the Being-question to be a definition? The question concerning the unity of Being in the multiplicity of its meanings is the spur which has urged Heidegger's thinking forward. This question alone, not simply that concerning the Being of beings, is the "Being-question" in Heidegger.

In his *Habilitationsschrift* the Being-question is silenced again because here Heidegger strives to reach a "metaphysical settlement" of issues such as how objective meaning could be valid, how *verum* could be convertible to *ens*, and how the manifold meaning of Being could be held unified. As demonstrated through a metaphysical and teleological exegesis, the fact that beings can be known, that meaning can have objective validity, and that consciousness of objects can be achieved calls for ascribing meaning to consciousness as what originally pertains to it. Accordingly a knowing consciousness bears upon its metaphysical origin; as finite and temporal, its ground is absolute Spirit. The latter is conceived as eternal residing in itself in a timeless manner and opposed to the "world" as the realm of temporality and transition. The valid meaning participates in the eternity of the Absolute and is thus static and self-subsisting, separate from the temporal actuality. More specifically, this meaning is separated from the course that the psychic phenomenon takes. Heidegger here operates on the basis of several presuppositions. He presumes that philosophy must have a metaphysical, teleological, and theological conclusion. Moreover, he assumes that the eternity of the Absolute, insofar as it is unchangeable, must be opposed to the temporality and changeability of the world. Finally, he presumes that meaning, insofar as it is true, must be intrinsically bound to an eternal, unchangeable, immutable self-subsisting Being.

In his university lectures after the First World War Heidegger abandons these metaphysical and speculative presuppositions. In accordance with the basic dictate of phenomenology, Heidegger tries in these lectures to acquire a "natural concept of the World." That is why he sets out from life in its facticity or factuality. As a living-in-the-world, life must be grasped from out of itself; no Transcendence, which would explain whatever there is, is to be presupposed by philosophizing, all the more so, since the God of philosophy is suspected to be no God of faith. Factually, life somehow understands itself already and it has meaning. When we consider the way it exists, meaning can no longer be conceived in keeping with the pattern in which eternity as immutable and statically self-subsisting contrasts with temporality as transitoriness. Rather, meaning is to be conceived as the motion of factical life which in its originality is history.

Life's facticity and historicity, Heidegger thinks, has been experienced in original Christian faith. In one of his earlier lectures, he quotes a statement of the Apostle Paul (from the first letter to the Thessalonians) as an example of how the facticity of life gets experienced. Concerning the return of Christ Paul states "the day of the Lord will come like a thief in the night." The return of Christ is neither to be dated chronologically nor can it be grasped with reference to a particular factor. Paul speaks only of its suddenness. The return of Christ pertains to the history of what life has fulfilled and this cannot be objectified. He who wishes to make available to himself, through chronological reckoning and via characterization of its factors, the unavailable and oncoming day of the Lord, deceives himself about the factuality of life. Paul states:

> When people say "there is peace and security" then sudden destruction will come upon them as travail comes upon a woman with child, and there will be no escape.

From the beginning Heidegger's thinking rests on the assumption that no thought which bars relation to the unavailable future escapes the doom by reckoning with time and appealing to available, imaginable, objective factors.

But is not "objectification" the nature of thought? Greek born and metaphysical, thought renders beings present in their Being. This thought is a "seeing" through which the Being of beings is placed in view. The Being of beings is taken here as being always in full view, as Being in a continuous presence. However, by going beyond beings to Being as continuous presence, does not thought fail to grasp factual life, which in its originality is the same as historicity? This indeed is Heidegger's opinion. In one of his earlier university lectures entitled "Augustine and Neo-Platonism" Heidegger discusses how Augustine falsified the original Christian experience of God, in light of which he lived and thought, by interpreting it with the aid of neo-Platonic, metaphysical concepts, as *fruitio Dei*. The *frui* in *fruitio Dei* as *beatitudo hominis* is a *praesto habere*, which is a having in view internally. Thus God's Being is conceived as Being continuously in view. This way, God could be enjoyed as peace, as imperturbability of heart and so easily forced out of disquietude of the factical historical life, thereby suppressing his vitality.

For metaphysics thought is a "seeing," and Being as Being in view is a continuous presence. Does not metaphysics thereby fail to reach the realm where beings may manifest their Being, namely, the realm of facticity and historicality? Does this forgotten realm pertain to metaphysical thinking? Should not the recognition that metaphysics conceives Being as continuous presence give rise to the crucial question whether by so conceiving Being metaphysics grasps it thoughtlessly in terms of a certain mode of time, the present? Does not continuous presence *(Anwesenheit)* mean being present *(gegenwärtig)*? Should not the question be raised whether metaphysics inconspicuously grasps Being in the horizon of Time? Inquiring in *this* fashion the question of Being undergoes a radical change requiring the following issues to be raised: Is it because the horizon of Time (within which Being is always already understood) is left unarticulated that the unity, diversified into the multiplicity of the kinds of meanings, of Being remains unconceived? Is it so because the horizon of Time *did* not seem worth inquiring into because Being as continuously present did not seem worth inquiring about?

If Being and the multiplicity of the kinds of its meaning are to be comprehended, we must consider as a problem the unity which diversifies into such a multiplicity. We must expressly inquire into the *sense* of Being which makes intelligible the multiplicity of its meaning. The sense of Being is the *ground* for the diversification into the multiplicity of the meaning of Being.

Insofar as Being is beyond any specification, its sense is to be comprehended as the *transcendental horizon* where Being as such is determined. By comprehending Being with a view to continuous presence and the present, by inconspicuously conceiving Being in light of Time, metaphysics itself leads to inquiring whether Time essentially pertains to the unity of Being as its sense, as the reason for its manifold kinds of meanings, and as its transcendental horizon.

In this fashion the question concerning the sense of Being becomes the question about Being and Time. On the first page of his work *Being and Time* Heidegger quotes the question raised in Plato's *Sophist* as to what we actually mean when we use the expression "being"? What do we mean, as Western Europeans, who think essentially in metaphysical terms, when we say "is"? We must consider the question, whether by conceiving Being as continuous presence we do not think of it quite naively, and forget Time that makes presence thinkable. Has not our ontological thinking, concerned as it is with Being of beings, forgotten its own ground, the sense of Being as it is pertaining to Time? Does this not make a fundamental ontology necessary which would establish, on its own ground, ontology as an inquiry into the Being of beings?

In the first two and only published divisions of *Being and Time* Heidegger expounds on the temporality of a being distinguished by its understanding of Being: he explicates man insofar as he understands Being, is Dasein. By setting out from the temporality and historicality of Dasein Heidegger wants to render Time thinkable as it pertains to the sense of Being. In the third division of *Being and Time*, setting out from the temporality of Dasein, Heidegger was to determine the temporal character of the transcendental horizon, a horizon wherein Being diversifies into multiple kinds of meaning. However, this attempt failed and the third division of *Being and Time* remained unpublished. The investigation of this work did not reach its goal.

After the failure of *Being and Time* Heidegger tried to make an inquiry into the transcendental horizon irrespective of the problem of temporality. What is transcendental about this horizon is Being as "transcendence pure and simple" which lies beyond the manifold Being of the various domains of beings as well as beyond the transcendental modes of Being. This Being may be grasped when Dasein transcends all beings, which in the difference of Being and beings is Dasein's accomplishment. This difference as the ontological difference, and with it the transcending of beings to Being, is the center of metaphysical thinking, the *meta* of metaphysics. But Heidegger does not accomplish only this metaphysical transcending of beings to Being. He goes further by directing his inquiry into the nature of the domain where such transcending is carried out.

In metaphysics, thought conceives the Being of beings; it conceives a being as it is in truth. If a being such as man is comprehended as a being, this being comes into its Being which is the openness of its truth. According to an unexamined assumption, metaphysics takes Being as what is present in the selfsame continuous presence. The assumption that Being is the same in all eternity, that man, at least in his Being and nature, has no history, appears to Heidegger worthy of being questioned. Therefore, he must ask: how is Being as Being to be thought of if it is not simply and thoroughly what is continuously present? What is Being itself, if I may be permitted to say so, in its "Being" or "nature"? With reference to what is Being understood at all "as Being"? How are we to think the sense of Being, or as Heidegger puts it now, its truth?

It is necessary that we differentiate precisely between the question of Being, as concerned with the truth of beings, and the question of the truth of Being itself. Nietzsche, for instance, responds to the question as to what constitutes the Being of beings, by holding that Being is life and life is will to power. In *Thus Spoke Zarathustra* he says:

> Where I found the living, there I found will to power; and even in the will of those who serve I found the will to be master.

The will to power is the Being of beings and the truth of beings. But what is this Being in its truth? Is this Being's distinguishing feature continuous presence so that beings are fundamentally will to power? Are beings will to power even when it may be impotence which is materialized as power? Or is this Being not at all distinguished by continuous presence, rather does the experience of the Being of beings, as will to power, indicate a definite and limited history and point out that Being has the feature of historicality?

Questions raised by Heidegger do not concern simply Being as the truth of beings. Rather his whole interrogative power is concentrated on the question of the truth of Being itself. Heidegger places the fundamental question of what Being itself is in its truth, before the guiding-question of metaphysics as to what is the Being of beings and its truth. He attempts to expose the unarticulated ground, which already supports metaphysics' concern with the Being of beings. In this manner Heidegger's inquiry into the truth of Being itself is a regress into the hidden ground of metaphysics.

In order to articulate the truth of Being as the ground of metaphysics, Heidegger inquires into the nature of truth. The existential analysis has already shown that truth is disclosed when Dasein succeeds in attaining its highest resolve. In this resolve Dasein must experience itself as nullified in a twofold manner. First, the fact that Dasein is and truth comes to pass, is beyond its power. Second, Dasein finds itself always in a situation, thereby allowing truth only a limited place for its openness. However, by releasing itself into this negativity, Dasein learns that it pertains to the nature of truth. The twofold negativity of Dasein, recurs as a twofold nothing in the nature of truth; to what is present, to truth as it opens up, is tied a twofold non-present and concealment. First, it remains hidden why truth opens up at all. Second, truth enters into a limited opening. In this manner, truth is a co-presence of unconcealment and concealment, a process of instituting and withholding of ground. Truth is ground; inasmuch as it stays away, it is non-ground (Abgrund). As the history of revealing and concealing, truth is the mystery of un-concealment or Aletheia.

Heidegger finds it likely that the negativity of that which does not come to presence and the negativity of concealment and refusal of truth to reveal its ground should be thought as pertaining to Being. The Naught as it pertains to Being is, of course, not the absolute nothing which can be only an ens rationis. Rather it is a no-thing in relation to beings. As it pertains to Being, no-thing is not a being, is not a property of a being and unavailable to any being. Being is the truth of beings only as a ground which is their non-ground. According to the twofold negativity, Being is not at all available: It is the unavailable history. As the truth of beings, Being is in its own truth the unavailable history; it is the event of Appropriation (Ereignis). In the event of Appropriation, the temporal and mittent-character of the truth of Being becomes manifest.

Since the years 1936-38, when Heidegger wrote the so-called "Contribution to Philosophy" (whose actual title is "On Appropriation"), he has thought of the truth of Being as Appropriation. The truth of Being needs Dasein as its instantaneous abode

(Augenblicksstätte); it appropriates itself to man by claiming man's comprehension of Being for its "here-there" *(Da)*. In the appropriatedness of Dasein to the truth of Being, the issue to be divulged, brought to maturity and settled, is the ontological difference as the difference between Being and beings. Divulging, maturing and settling this issue, beings may be placed in their identity and Being. Being and man as a being that understands it belong to each other in the "here-there" of Being. In its truth man and Being belong to each other in such a way that Dasein can never have at its disposal the truth of Being, but must adapt itself to the history of this truth. The truth of Being is not a ground that can be determined and secured; it is a ground which institutes historically, is unavailable, and is non-ground.

If the truth of Being as Appropriation is the unavailable history, then submersion in this truth and regress into the ground of metaphysics can also be history and thereby decision. Heidegger encounters in Nietzsche the decisive thinker, insofar as Nietzsche draws ultimate consequences from thought's metaphysical beginning. In this beginning, Being is taken as a continuous present, without inquiring into either the presence of this present, the horizon of Time for Being, or into the truth of Being itself. If Being is thought according to its "nature" as continuous presence, if it is taken as what is always present, then Being and beings become available to man provided that he utilizes his thinking decisively. Perhaps the reason why Being is thought only as what is continuously present is that in this way it may be made available to thinking. No sooner does man make himself free to wish and to be able to dispose over beings than he renders beings present by representing them in their Being. Modern times begin when the Being of beings is made representative of objects. When conceived as representativeness, Being conceals the Will as what is essential for summoning beings and placing them before us. That is why Nietzsche draws only the final consequences of the thinking commenced in metaphysics and transformed in modern times, when he conceives the Being of beings as the Will, which, in all its representations and presentations, wills nothing but itself and is Will to power. The Being of beings, which is continuously present in all beings as their underlying addendum or *subjectum*, is now subjectivity unconditionally attaining its perfection as the will to power. By willing itself to return eternally, thus making itself permanent in continuous presence, this subjectivity becomes what it is, namely, the underlying addendum as that which is constantly present. Thus it becomes grossly obvious that the desire to have Being at one's disposal always prevails when, at least inconspicuously, the Being of beings is conceived as constant present. In this "completion" of metaphysics, it becomes impossible to regard continuous presence as a problem; likewise it becomes impossible to make an inquiry into the temporal horizon of presence as an inquiry into the truth of Being. Continuous presence cannot be taken as a problem where, by willing the eternal recurrence, will obtains for itself continuous presence. Perhaps Nietzsche, who pushes metaphysics to the extreme, can initiate an inquiry into metaphysics as a whole, so that we may grasp metaphysics as a history which in its inception forgets the truth of Being as its supporting ground, and heads fatally towards its end. Seen in this manner metaphysics would not be the history of truth in general but the history of thinking Being as the truth of beings, which according to the way it begins, has to leave unthought the truth of Being.

Metaphysics thinks being in its Being. It inquires into the Being of beings of various spheres. It conceives being generally according to the main features of its Being by differentiating for example quiddity (what-being) from existence (that-being). In the doctrine of transcendentals, metaphysics presents various aspects of Being which itself offers, for instance, Being as Being-true. Metaphysics reflects upon the Being of beings in its entirety by conceiving this Being as Platonic idea, as representativeness of objects in modern times, and finally as will to power. Then metaphysics is the doctrine of the Being of beings--it is ontology. This ontology takes as a matter of course continuous presence as the basic feature of Being. Ontology presupposes that beings can be grounded upon Being as what is continuously present and also available. Being itself needs also to be grounded upon something for it to be what is continously present. Thus metaphysics searches for a being which can expressly fulfill the requirement to be continuously present. It finds this being in the self-subsisting divine being, in THEION. Therefore metaphysics is not only ontology as process of grounding beings in Being but also theology, as process of grounding Being in the highest being, in THEION. Because metaphysics is fundamentally concerned with the reasoning for and the establishment of ground and is a -logy, it is onto-theo-logy. However, having established itself as Being of beings, will to power does not let itself be grounded upon another being. It obtains the process of establishing ground by willing itself eternally recurrent and constantly present. Will to power sets itself up as its own foundation, thus disposing over itself as the ultimate ground of whatever there is. It corresponds to the chief onto-theo-logical feature of metaphysical thinking that will to power which wills itself eternally is named by Nietzsche after a God, Dionysus.

The greatness of metaphysics consists in its conceiving the Being of beings and initiating an ontology towards a self-subsisting divine being. The failure of metaphysics consists in positing Being initially, in an unexamined manner, as constant presence. This view gives thereby the character of availability to Being and consequently turns thought, as it initiates the grounding process, into the ability to dispose over ground and the ultimate ground. Thinking in this fashion, man is expelled from his nature in its relation to the historical and unavailable future. Consequently inexhaustible nature, shining forth and retreating in its abundance, becomes an object of representation, subsisting only to be called forth. Accordingly man becomes, in Nietzsche's words, "murderer of God;" the divinity of the divine as continuously present and available is not left untouched when it is placed face to face with man at his direction. By thinking this way, metaphysics distorts being and its Being.

In the same vein metaphysics is unable to trace such ontological differentiations as existence (that-being) and quiddity (what-being) back to their original descent and limitation. Heidegger tries to show that the difference between existence and quiddity results from Being coming to prominence as what is continuously present: what is present, as what exists, and its outward appearance, as its whatness. If the equation of Being with continuous present seems worthy for inquiry then it must also seem worthwhile to ask how far does the range of beings reach to which the difference of existence (that-being) and quiddity (what-being) is applicable. That the sustaining force of this difference is restricted, that such a difference cannot be applied, e.g., to the factical, historical existence of man, that is what Heidegger's thinking started from.

According to a basic assumption of metaphysical thinking, the doctrine of transcendentals, for instance, held Being-true as convertible to a continuously present Being, as the openness of what is always present and available. Thus truth was placed at man's disposal and if it was not altogehter at his disposal, then it was at the disposal of a God who was conceived in a certain way. In Plato's *Republic* (597) it is said that God created for every thing, *one* idea, one *lasting* Being, because he did *not* want to be the creator of one single unique thing but of a true being whose

Being is continuously present. Are we supposed to think of God this way?

By conceiving Being as constantly present, metaphysics occupies a definite place in the history of the truth of Being. However, not having known the truth of Being as the history which assigns metaphysics to its place, it can carry this place in itself only as a forgotten one. Conceiving Being as the truth of beings and not reflecting further upon what it does, metaphysics puts forth Being as the constant present. Thus neither the truth of Being, nor an inquiry into this, appears in metaphysics. The truth of Being fails to become the event of Appropriation. Individual metaphysical thinkers also do not reflect on the place they occupy in the vicinity of metaphysics when they take Being, continuously present, as idea, as representativeness of objects or as will to power. A discussion of the place of metaphysics and places occupied by individual thinkers can be accomplished by a thinking which thinks Being in its history as the absence of the truth of Being and as a possible Appropriation of this truth.

Metaphysical thinkers pass on their own truth as the sole and ultimate truth. To Nietzsche, for instance, the doctrine of will to power is, to be sure, an "experiment of the knowers" but this experimentation does not take the will to power as an explication of the Being of beings which is consistently modern and historically located. Rather will to power is regarded as a recognition of an ultimate fact. Nietzsche, of course, notes that will to power has not always been cognized and recognized as an ultimate fact. However, he interprets the behaviour of Platonists and Christians, who put a God above them, as an impotency, which is basically nothing other than will to power. The metaphysical vision of a permanent Being as the truth of beings wants to offer nothing less than a final ground which sustains processes of grounding. This vision wants to be a fundamental insight and a fundamental teaching.

Eighty years ago Nietzsche stated:

There will be a time when domination of the earth will be contested through fights waged in the name of *Philosophy's fundamental teachings*.

Fighting for domination of earth means a struggle to put earth as the entirety of available beings at one's disposal, and to find out the humanity capable to take over such an arrangement. Indeed, the time of such fighting has arrived, and it is waged with means about which Nietzsche could not even dream. However, is it waged in the name of philosophy's fundamental teachings? This fight is waged most vigorously in the name of philosophy's fundamental teachings in places where perhaps philosophy and metaphysics are out of the question, where metaphysical inclination to offer and secure a final ground for things is pushed to extremes; where without slightest reflection upon what is being done, a highest principle, such as a permanently present Being is set above history as a goal to master it. Those who wage this struggle could learn that the permanent ideological goals lose their credibility and are nullified. To know this nullification, to be profoundly touched by nihilism, results in crying for another goal to be set, for a new "affirmation" and a new courage "to be."

But perhaps, Heidegger muses, nihilism can offer us still another opportunity. A profound experience of nihilism can possibly call upon us not to take the Naught as exclusively identical with the nullification of this or that goal or with opposition to the promotion of a certain entity to the rank of the highest being, but to think of this Naught as intrinsically belonging to Being itself. Thus Naught could come to be known, Heidegger maintains, as "the veil of Being," a veil that occasionally shows us Being by withdrawing it from us, thereby making it impossible greedily to reach for Being as continuously present, as available and as the ultimate ground of all beings. If the Naught which owns man's Being as Being-unto-death is thought of as intrinsically belonging to Being, then Being would show itself in its truth as the unavailable event of Appropriation.

The question of the truth of Being articulates the ground of metaphysical thinking. But insofar as Heidegger is concerned this question is itself no longer a metaphysical one. It is the question forgotten in "metaphysics." It is more likely, Heidegger holds, that a trace of the truth of Being could be found in pre-metaphysical, pre-Platonic or pre-Socratic early Greek thinking. Be this as it may, we do not receive properly Heidegger's quest for the truth of Being when we grasp it as a traditional metaphysical question. If we maintain that Heidegger conceives Being as event of Appropriation, or to state it less precisely, as history or Time, if we put the designation "Being as event of Appropriation" on the same level with the metaphysical assumption about Being as continuous presence, then the whole thing is misunderstood. Heidegger does not replace metaphysical assumptions about Being as constant present with another designation. Rather his inquiry is directed at the place in view of which Being can be comprehended at all. He raises the question as to what is hidden in "as," in the expression "Being as Being": he inquires into the truth of Being itself.

When the equation of Being with constant present becomes problematical and therefore explicitly requires an inquiry into the truth of Being itself, it is in no way denied that within limited areas of beings their Being can be thoroughly a continuous present. For example, by having the characteristics of continuous presence the condition of symmetry in a mathematical equation makes it possible for us to return always to the symmetry of the equation. Moreover, problems such as that of natural right, of archetypes or prototypes, the problem of a "general metaphysics" are problems related to the definitely legitimate question as to whether being, in limited areas, is not the same as what is continuously present. The thesis that Being of beings is not continuously present, but identical with a thoroughgoing temporal transitoriness, would be refuted for example by the demonstrable Being of mathematical objects. But Heidegger has never put forward such a thesis. He has denied only that the Being of beings is exclusively of the nature of continuous presence by directing his inquiry into the area where the Being of beings is determinable at all. We must sharply distinguish what Heidegger calls the "history" of Being and its truth as "event of Appropriation" from what is usually called history which is the region of beings other than nature. What is nature and what is history, in their Being, can be thought adequately following the experience of the truth of Being.

Initially Heidegger referred to this truth as "Being itself." Thus around 1936 he speaks of Being as event of Appropriation, meaning by "Being," Being itself, its truth, writing it at times with an antiquated spelling to differentiate it from the Being of beings. By articulating the truth of Being (literally bringing it to language) as the event of Appropriation, he gives up the name of Being. The word Being, then is returned to metaphysics, since by this word it conceives the Being of beings. Now Being as the Being of beings can only be thought *from out* of the truth as the event of Appropriation. Thus the semblance of "event of Appropriation" with a transcendental metaphysical determination of Being disappears. Accordingly to speak "of Being" already presents a metaphysical hypostatization. Whereas it is originally appropriated truth that teaches us how to think the Being of beings in its entirety and how to determine the Being of each realm of beings, such hypostatization wants us to believe that we could encompass

Being and grasp it.

In Heidegger's later thought the basic word is no longer "Being" but "truth." The question which he now develops is concerned with what structurally has joined together to set up truth (*Baugefüge*). First he shows in art how truth comes to be appropriated. Truth as put to work in art is the lasting and enduring origin, is a rising forth on and on forever, which like a source conceals itself in its inexhaustibility. Truth makes a being conspicuous in its Being or "truth" by refusing and concealing itself. Thus as the truth of beings, truth not only lays open for knowledge, it also closes off and is concealment. Laying open and being manifest Heidegger calls "world," closing off he terms earth. Truth in its originality is the strife of world and earth. The work of art brings a being to its truth, it turns for us into world the self-secluding earth, it lets us reside in non-ground which has no abode and withdraws. In all this, however, art lets earth remain earth, self-secluded, it makes possible not only manifestness and openness to shine through things but also the inexhaustibility and inscrutability of those depths which never entirely are surrendered.

The very heart of truth is mystery which is neither the same as what is simply screened off from view nor an enigma that once will find its final solution, but the *persisting* mystery. This mystery grants occasionally an opening by receding into concealment. Man does not have the God-like intelligence to resolve the mystery of truth. Rather, as mortal he is permanently assigned to this mystery. Finite and mortal, man does not seek the manifestness of things alone but also is concerned with what is wholesome and hale and what is unwholesome. He wants to find out whether existing things reside in their truth or dwindle in an empty nothingness. If man knows himself to be a mortal, if he comes to know truth as the unavailable mystery which only at times opens up, then for him truth becomes the expanse of the holy and the divine--those who are wholly other than man yet address him and "need" him. In Heidegger's view it is Hölderlin who experiences truth in its originality as the holy, as the expanse for the divine's appearance, and this at a time when the onto-theo-logical trend toward producing and securing an ultimate ground for beings had "murdered" God and deprived the divine of its divinity. If Heidegger draws upon Hölderlin's poetic utterance on God and the holy, this must not be taken to mean that he substitutes another expression of the holy with that of Hölderlin. As little as *Being and Time* becomes "theological" by references to Augustine,

Luther and Kierkegaard so too Heidegger's later thought is "mythological" in its relation to Hölderlin. Apart from the fact that Hölderlin's poetic expression is not a mythology to which reference could be made, a thoughtful relationship to him can assimilate Hölderlin's experience only as a *question* about how contemporary man, and specifically man of the future, may come to know the divine's claim.

What in the first beginning of our history remained forgotten as the truth of Being, Heidegger seeks to ascertain as that which makes possible "another beginning" of our history. This truth, he thinks, is set up and structured as the strife of earth and world, and as the interrelation of the divine and mortals. Later Heidegger uses for world as that with which earth is in strife, the expression "heaven" as the open domain which pertains to the self-secluding earth. This enables him to call "world" the *entire* setup and structure of truth. World in this sense must, of course, be sharply differentiated from the metaphysical concept of world as the totality of entities. It must also be distinguished from the historico-anthropological concept of world as a wholeness which consists of historically projected meaningful relations. World as the interrelation of earth, heaven, divine and mortals is the Appropriation of truth. When it comes to be known as granting the wholesome and the holy, truth bestows world as being indigenous, as a while of being-at-home, as an abode.

Let us look again at the course that Heidegger's thinking has taken by concentrating on the question whether this course presents a thinking way and if so let us ask in what does the "method" of his thinking consists.

In the beginning of his thinking way Heidegger took up the metaphysical problem of the Being of beings. He sought the unity of Being which diversifies into the multiplicity of the kinds of its meaning. The inquiry into the unity of the multiplicity of Being underwent a radical transformation when Heidegger subjected the doctrine of Being in metaphysics to the question whether this ontology satisfies the facticity and historicity of life. Does metaphysics not hold within it the realm of facticity and historicity as the forgotten realm, inasmuch as it conceives Being as continuous present without reflecting on the horizon of Time wherein presence and presentness are to be thought? *Being and Time* is a split and disunited endeavor, which had to fail precisely *because* of its split and disunited character and no other reason. On the one hand this work is on its way to thinking the truth of Being, on the other hand it still speaks an inadequate "metaphysical" language. In a metaphysical and modern manner it sets out from transcendental subjectivity as *fundamentum inconcussum veritatis* and calls the Being of this subjectivity "existence," an expression taken from modern metaphysics of will. In its way of thinking, to be sure, *Being and Time* opposes modern transcendental philosophy and metaphysics since it represents the Being of existence as finite temporality. However, there is no transition from temporality of existence to the character of time inherent in the sense of Being because existence is not yet satisfactorily differentiated from modern subjectivity by historical reflection. In accordance with its very nature modern subjectivity screens from its own view the truth of Being. It seeks to be its own sustenance and, therefore, cannot fit into the coming-to-pass of the unavailable truth. This subjectivity is not Dasein as Being's here-there *(Da)* and its instantaneous abode. Furthermore, *Being and Time* takes the sense of Being as the ground on which everything should be placed. To be sure, the historicality of thought is to be brought out by "destruction of the history of ontology." However, this destruction does not allow thought to lend itself to the unavailable history of truth; instead it is to arrive at what is ultimate but contorted. By seeking to prepare a foundation for every ontological inquiry, *Being and Time* still follows the metaphysical pattern of reaching for a final ground. Inasmuch as it comes to be known as Appropriation or history which forever comes-to-pass in a sudden way *(jdh)*, how could the truth of Being be a firm foundation, a producible ultimate ground? At the most it could be called non-ground, but taken in its own right it cannot be called "ground" at all.

After the failure of *Being and Time* Heidegger sought to turn his thinking to the truth of Being which evades the process of grounding. This turning, also called "reversal," is a turning from Dasein to Being. The talk about the turning misconstrues everything as long as one does not see that this turning has never been carried out and could never be accomplished as planned in *Being and Time*, namely, as the systematic transition from the analysis of existence to the sense of Being. Turning is achievable only as a turn about *(Umkehr)* through which thought resigns from supplying a final ground in order to sustain and secure itself. This turn-about must undo the history of metaphysics and in *this* sense alone it is the way back into the ground of metaphysics. For Heidegger this retreat is the same as passing decisions on self-assertions of man and the search for a ground. Let us recall that Heidegger in his Chancellor's Address of 1933 called Prometheus the first philosopher because by *stealing* the fire from heaven he asserted himself. The metaphysical trend to reach out for an ultimate ground is maintained in the fundamental ontology which Heidegger, using Kant's words, called "a metaphysics of metaphysics." This expression pretends that the inquiry into the nature of man can "outbid" and justify the metaphysical inquiry into the Being of beings. To state it in terms other than those of modern and transcendental philosophy we can say: the expression "a metaphysics of metaphysics" pretends that a metaphysical inquiry into the Being of beings can be brought to bear upon itself so that an inquiry into the "Being" of Being (an inquiry into the "nature" of metaphysics) would be made possible. However, no sooner is the truth of Being itself articulated as the non-ground of the event of Appropriation than this pretension disappears. Metaphysics cannot be "overcome" by being subjected to the process of grounding, it cannot be done away with by reaching for something higher than metaphysics. Rather, by yielding to the truth of Being, thought must give up the metaphysical will to produce and secure an ultimate ground. Thought cannot overcome *(überwinden)* metaphysics, it must try to incorporate *(verwinden)* it.

The question "What is Metaphysics?" as Heidegger points out in *Zur Seinsfrage* (1956, p. 37) stabs itself in the heart not in order to terminate the life of thought but for its transformed continuation. When concerned with a "what" or an "essence" the question "What is Metaphysics?" is itself put metaphysically. This *metaphysical* way of putting the question must be given up if the truth of Being is to be experienced as a history which, though sustaining, is unknown to metaphysics. As history, Heidegger subjects metaphysics to a decisive proceeding which through its death leads to a new life.

In the history of metaphysics Hegel was the first to learn thought's breaking out in death as history. It is reported that in the winter semester of 1805/06, when Hegel first lectured on the history of philosophy, during one evening by lamplight, he took one representative of thought after the other, dealt with them, and left them behind. Finally Hegel took up Schelling's thought, dealt with it, and left it behind as inadequate. Then a rather aged student, jumping to his feet, declared in dismay that this is death and the end of everything. A lively dispute followed, until one of the students of Hegel stated that this of course is death and it must be death, but there is life in this death, a life which purifies itself through death to unfold itself ever more magnificently. In the *Phenomenology of Spirit,* Hegel considered the prime task of thought to be the passage of it through death to purify itself, which is thought's turning against itself in order to find back a way to itself as history. After having thought of original truth as history Heidegger had to come to grips with Hegel's *Phenomenology.* Heidegger's closeness to Hegel is obvious, but more crucial is what separates both thinkers. For Hegel life that originates from death is the life of the Absolute. The passage through death and through the entirety of history is by Hegel cancelled, preserved and elevated *(aufgehoben)* in an absolute Being-with-itself to which nothing is not manifest. By including history in the life of the Absolute Hegel made possible that virulent metaphysical presumptuousness about the Absolute which after him has determined the world's history. Man, due to his Being-unto-death, is, according to Heidegger, assigned to the unsurpassable mystery of there being something existing at all and of coming-to pass of truth to be appropriated, and herein lies Heidegger's decisive difference from Hegel. A thinking transformed by death, from its own location seeks to be incorporated into the unavailable event of truth. Such thinking opens for man an access to truth only through *questioning:* it is unable to provide him with an Absolute, much less to tell him concretely the truth of what to him and his position is wholesome and unwholesome.

The fundamental movement in Hegel's thought is to "cancel, preserve, and elevate" *(aufheben)* everything in the Being-with-itself of the Absolute. In contrast, the fundamental movement of Heidegger's thought is, in his own words, "the step backward." He takes up the metaphysical question of the Being of beings in order to retreat into the truth of Being, to articulate it as what has been left unthought. The quest for the truth of Being must do away with even *the* pretension as though it is potentially a metaphysical inquiry. A discussion of Being with respect to its truth, a discussion of the *topos* of Being *(Seinstopik)* as I would like to call it, is itself no longer a "metaphysical" discussion. By discussing the truth of Being as what is left unthought in metaphysics, a discussion of the *topos* of Being abandons the region *(Ort)* of metaphysics where the truth of Being fails to appear, in favor of a region where this truth can become Appropriation. Thus "discussion" in this novel sense presents a pathway to be ventured upon not by abandoning one position in favor of another but by means of a discussion which returns to the unthought of what already has been reflected upon. Such a discussion which articulates the unthought arrives from one region into another; remaining on its own place it incorporates itself into the history of truth. The course that such a discussion takes is not simply Heidegger's course of thought, but it should be nothing other than a retaking of the course of European thought by retreating into the beginning of our history. Retreating into this beginning, the unthought in it can be discussed, thereby restoring to this beginning its incipient character so that it can become "another beginning."

By means of a discussion of Being, as a discussion of its *topos*, Heidegger evades metaphysical thinking in that he articulates (literally brings to language) what is left unthought in this thinking. In this manner, he refers metaphysical thinking back to its region and its limits. By determining the relationship of metaphysics to a discussion of the *topos* of Being, I would have arrived at the end of my essay. I would like, however, to add briefly a few suggestions which may clarify a bit what discussion as *Erörterung* is.

First I would like to differentiate discussion from explanation and clarification. Explanation means to reduce a being to another being or to "Being" as its ground. "There is nothing without a reason or ground" is a principle which makes it possible to explain and control beings. In the nineteenth century explanation became the method of sciences. It not only dominated sciences of nature but human sciences as well. Religion for instance was reduced, rationalistically and positivistically to psychological conditions such as man's fear, and was considered explained. In a less coarse explanation, religion was traced back to the history of the individual psyche and the history of the psyche of the people. At times explanation arrived at such ultimate principles as "extension" or "consciousness." As the reference to both Cartesian substances indicates, presumably through explanation as a method, metaphysics reached out to grasp an ultimate ground. This presumption, of course, should not be taken to mean that the way metaphysics is instituted ontologically has to be branded as ontical explanation. It should rather prompt the question whether the way metaphysics is instituted ontologically distinguishes itself adequately from ontical explanation to exclude any misunderstanding, and whether its coming close to an ontical explanation was due to not thinking the truth of Being itself.

It amounts to a liberation of thinking that at the beginning of our century ontology and phenomenology opposed explaining by pointing out that, for instance, concerning religious phenomena, nothing is agreed upon when religion is construed as an expression of a historical humanity, when it is explained in view of something else. Opposed to explaining, we have a thinking whose essential feature can be designated as clarification. Accordingly, clarification means: to let something abide in the purity of its nature by refraining from any hasty explanation. The demand, "back to things themselves" signifies, not to explain logic psychologically, not to quickly take religion or poetry as an expression of a people's mind, but by means of an unprejudiced description of things to let them be seen primarily as what they are. The danger of clarification, however, consists in taking, once again, consistent with the trend of metaphysical thinking, purity of the essence as what is *continuously* present and having it reside in a supercelestial place. This danger lies in assuming a realm of logical laws "in-themselves" *(an sich),* values "in-themselves," generally essences in them-selves, or to use a modern expression, to reduce everything to a transcendental ego which lies beyond any demonstrable ego. Consequently the problem arises of how concrete history can be related to ethical values themselves, religion itself, poetry itself?

Heidegger now seeks to show that no relation needs to be established here at all, as long as we do not forget that the Being of beings or nature of things accrues in the domain of the truth of Being itself, for which historical man is "used." Initially Heidegger referred to the domain of the truth of Being through man's historical understanding of Being. But "understanding" as a guiding-term remained ambiguous. It seemed as if understanding in an historical sense would be the same as self-understanding *of* humanity used as a ground for explaining everything. The term "discussion" as *Erörterung*[1] articulates more adequately what is actually the issue in Heidegger's thought: the truth of Being itself which as Appropriation always situates our inquiry into the Being of beings, in its site and place. Being as thought through the truth of Being is itself a situated Being *(beorted)*, it is from the first finite and historical, even though in limited areas there may be a being always enduring, to which we can return, irrespective of our historical situatedness.

In this sense, *Erörterung* is the logos of Heidegger's thinking. Perhaps we use language "authentically" when we use it as a means for *Erörterung*. Heidegger, differentiating poetical speaking from speaking proper to thinking, states succinctly: the poet names the Holy, the thinker utters Being. What the poet does, to wit, responding to the claim of the Holy, and as Heidegger joining Hölderlin says, naming the Holy and the divine, is what the thinker cannot pretend to do. The thinker utters Being, which is to say that he discusses the utterance of the Being of beings through the truth of Being, in order to direct man to the history of truth by means of *questioning*. In this truth and what structurally has joined together in its setup as world, there is no onto-theo-logically conceivable ultimate ground. Proportioning the claim of the divine, there is always a measure, but the thinker at least cannot possess this measure as an ultimate one. That is why he has no right to construct an all-encompassing metaphysical system. Not only this or that system, but the idea of system as such has become questionable. Since thought does not possess an ultimate measure it cannot substantiate the "world" by wishing to legitimize it or correct it, as it is displayed by physics and biology. Of course thought can discuss, over and over, the unexamined presumptions that have gone into physics or biology. Heidegger, for instance, tried to determine the status of classical physics by showing that originating as it does from a conception of Time limited to a particular image of world, a number of assumptions have gone into this physics as pure science that do not belong there, a view after all substantiated in its own way by modern physics. A discussion of statements made by individual philosophical disciplines or those made by certain sciences has to set out from unexamined philosophical or other implications accepted by them, such as the concept of infinity in mathematics, the basic concepts of aesthetics, lastly the concept of aesthetics itself, etc.

The crucial question is not whether discussions like these are carried out but how discussion as *Erörterung* can be legitimized as a basic method of thinking. Is, in its primordiality, truth an event in need of discussion, is it a way leading to various regions? If so, can this be demonstrated? If it were possible to demonstrate that this is the case, then we would have already transcended discussion. Situated as it is in its region discussion incorporates itself into the coming-to-pass of truth, but it cannot encompass the truth as a way that leads to various regions. If it is the case that such a way is experienced in its own region as different from another region, and that it cannot be substantiated in its entirety, then does not this rule out the actual legitimization of the discussion? Or is the question of such a legitimization an ill posed question? It can indeed be shown that Heidegger arrives at a thinking which discusses the utterance of the truth of Being, since according to his experience the process of grounding and explaining in metaphysics has reached its end. It is this ending in metaphysics which refers us to a thinking which discusses the utterance of the truth of Being. What other kind of unexamined presumptions determine the course of such thinking will be found only by one who ventures such a thinking. As an "absolute method" this thinking cannot be "legitimized."

Why do we call a thinking that discusses the utterance of the truth of Being with the old title of *topics?* It has been stated with respect to this thought's earlier attempts that Heidegger follows a "hermeneutical logic," a logic not oriented to things that are always handy but directed at historical life as it is capable of self-interpretation and at the practicality of such life. Of course "practicality" is the magic word which after Hegel's time has been the focus of lively interrogation. However, in viewing our practicality we view ourselves, and in viewing ourselves, we are easily prevented from experiencing the unavailable history of truth. Later Heidegger situated his thinking close to poetizing and in the vicinity of the aphoristic thinking of the "pre-Socratics." If we search for a thinking which is aware of the way it leaves behind, as a correspondence to *Erörterung*, then we come across *topics*. (Of course, this relation to *topics* is not established by Heidegger himself.)

Topics is the *ars inveniendi*, the art of finding arguments and basic concepts for a dialogue about something. It serves dialogue and its art, "dialectic." Topical and dialectical thinking does not offer truth as what is convertible to a continuously present Being, it offers only what is probable, *endoxon* or *verisimile*. Granted that shining forth and concealing pertain essentially to truth, granted that truth in its primordiality is what seems to be true *(Wahr-Schein)*, then topical, dialectical thinking must not be devalued because it imparts "only" what is probable. Topical thinking was regarded once as significant. In the tradition of exegesis of Aristotle, for example his writings on *Topics* were put together with those dealing with categories, since the latter are first to be found through some kind of discussion of *topos* of Being. For the most part, however, topical-dialectical thinking is forced out of the "first science" and forced into rhetoric, into humanism (which opposed the Scholastic philosophy), into jurisprudence and into theology. For the sake of man's historicality, Giambattista Vico, whose background was rhetoric, tried to justify the "old" topical method against the "new critical," straight systematic method of Cartesianism. To the student of law *Topics* was inevitable and out of his own practice he distrusted the straight systematic method. He knew: *omnis definitio in iure civili periculosa.*[2] He had to interpret the written law, which is to say that he had to adjust it to the historical change. Theological dogmatics, having to articulate a unique historical claim, are opposed to rigorous systematization. Far into the modern era dogmatics were treated topically and brought under the title "Loci Communes."

Does not tradition give an indication for articulating an original thinking centered at the very heart of metaphysics and time and again suppressed, which is a topical thinking oriented to "regions"? Be that as it may, for us here the question is, in what shaping mold should we place Heidegger's work as presented in recent publications of his numerous talks, lecture courses, essays, papers and letters? These endeavors of Heidegger's thought present one single great discussion (in the sense of *Erörterung*), capable of bringing about an utterance *(Sagen)* in the region we occupy as another region. In this discussion Heidegger proceeds in such a way as to concentrate on individual guiding-terms and guiding-propositions, that is, to focus on *topoi,* as I would like to call it, by using the terminology of *Topics.* Therefore, a thinking given wholly to such a discussion could perhaps be called a "topology." In this case topology means a saying *(legein)* of the region or site *(topos)* of the truth, a determination of the region which unfolds as places of gathering, and gathering-together *(logos)* of guiding-terms *(topoi)* of European thought and in this way a gathering of the basic terms of one's own thinking. *Topoi* are guiding-terms like *aletheia, idea,* and guiding propositions such as "principle of ground," "poetically man dwells," also grammatical forms like subject-predicate relation in sentences which has determined our thinking decisively. Each of these *topoi* points at one region of truth.

Topological thinking tries to achieve something very simple: it calls our attention to presumptions hidden in concepts we use, it seeks to speak language meditatively by asking us to keep in mind that in speaking as we do, namely from our site, we get incorporated in the coming-to-pass of truth. As the topology of Being this thinking is a meditation on what we Europeans actually mean when we say "is." Subsequent to the failure of *Being and Time* Heidegger devoted his thinking more decisively to the simplicity of such a meditation. Should not such a meditation be granted a saying of Goethe that Heidegger mentions at the end of his letter to Ernst Jünger, which was the conclusion of an essay concerned with adding to Jünger's description or topography of nihilism, a topology by indicating the region of speaking and describing? Goethe states:

> If someone takes word and expression as sacred testimonials and does not wish to use them, like currency and paper money, in quick and immediate circulation but wants to have them exchanged in intellectual transaction and transformation as true equivalents, then one cannot blame him for having drawn attention to the fact that traditional expressions, with which no one finds anything wrong, nevertheless exert a damaging influence, obscure the view, distort the concept and give a false direction to the entire fields of research.

Translated by Parvis Emad.

NOTES

1. To understand more fully what Pöggeler has in mind when focusing on *Erörterung,* it may be helpful to recall Heidegger's own treatment of this term. In his essay on the poetry of G. Trakl he calls *Erörterung* "pointing toward a place, showing a place and heeding it." Cf. *Unterwegs zur Sprache,* p. 37. Neske, 1960, English trans. *On the Way to Language,* p. 159, Harper & Row, 1971.

2. "Every definition in civil law is dangerous."

BIBLIOGRAPHY

This bibliography is limited to the publications cited in this book. For an extensive bibliography of works by Heidegger and of publications on his thinking I refer the reader to the one prepared by Hans-Martin Sass (*Martin Heidegger: Bibliography*. Bowling Green: Philosophy Documentation Center, 1982).

This bibliography is divided into three sections:

   I. Works by Heidegger in chronological order with English translations.
  II. Books and articles on Heidegger's philosophy.
 III. Collections of essays on Heidegger's thought.

I. WORKS BY HEIDEGGER IN CHRONOLOGICAL ORDER WITH ENGLISH TRANSLATIONS

"Das Realitätsproblem in der modernen Philosophie." In *Philosophisches Jahrbuch* 25 (1912), 353-63. (Cf. FS, pp. 1-15.)

"Neuere Forschungen über Logik." In *Literarische Rundschau für das katholische Deutschland* 38 (1912), 465-72, 517-24, 565-70. (Cf. FS, pp. 17-43.)

Die Lehre vom Urteil im Psychologismus. Ein kritisch-positiver Beitrag zur Logik. Leipzig: Barth, 1914. (Cf. FS, pp. 59-188.)

Die Kategorien- und Bedeutungslehre des Duns Scotus. Tübingen: Mohr, 1916. (Cf. FS, pp. 189-411.)

"Der Zeitbegriff in der Geschichtswissenschaft." In *Zeitschrift für Philosophie und philosophische Kritik* 161 (1916), 173-88. (Cf. FS, pp. 413-33.)

    "The Concept of Time in the Science of History." Translated by H.S. Taylor and H. W. Uffelmann. *The Journal of the British Society for Phenomenology* 9(1978), 3-10.

Sein und Zeit (1927) Tübingen: Niemeyer, 1953.

    Being and Time, trans. John Macquarrie and Edward Robinson. London: SCM Press, 1962.

Kant und das Problem der Metaphysik (1927). Frankfurt: Kloster-
mann, 1951.
  Kant and the Problem of Metaphysics, trans. James S.
  Churchill. Bloomington: Indiana University Press, 1962.
Vom Wesen des Grundes (1928). Frankfurt: Klostermann, 1928.
  The Essence of Reasons, trans. Terrence Malick. North-
  western University Press, 1969.
Was ist Metaphysik? Frankfurt: Klostermann, 1955.
  (Postscript added to 4th edition in 1943; Introduction added to
  5th edition in 1949)
    What is Metaphysics? (1929) trans. David Farrell Krell in
    Martin Heidegger, *Basic Writings*, ed. David Farrell Krell.
    New York: Harper & Row, 1977, 95–116.
    "Postscript" to *What is Metaphysics?*, trans. R. F. C. Hull
    and Alan Crick in Werner Brock, *Existence and Being*.
    Chicago: Regnery, 1949, 349–361.
    "Introduction" to *What is Metaphysics?*, trans. W. Kaufmann,
    in W. Kaufmann, ed., *Existentialism from Dostoevsky to
    Sartre*. New York: New American Library, 1975, 265–279.
Vom Wesen der Wahrheit (1930, 1943). Frankfurt: Klostermann,
  1961.
    On the Essence of Truth, trans. John Sallis, in *Basic Writ-
    ings*, 117–141.
Einführung in die Metaphysik (1935). Tübingen: Niemeyer, 1953.
    An Introduction to Metaphysics, trans. Ralph Manheim. New
    Haven: Yale University Press, 1959.
Holzwege (1936–1946). Frankfurt: Klostermann, 1950.
    The book contains six essays whose English translations
    appeared in different collections:
  "Der Ursprung des Kunstwerkes," *Holzwege*, 1–68.
    "The Origin of the Work of Art," in Martin Heidegger,
    Poetry, Language, Thought, trans. Albert Hofstadter. New
    York: Harper and Row, 1971, 7–87.
  "Die Zeit des Weltbildes," *Holzwege*, 69–104.
    "The Age of the World Picture," in Martin Heidegger, *The
    Question Concerning Technology and Other Essays*, trans.
    William Lovitt. New York: Harper & Row, 1977, 115–154.
  "Hegels Begriff der Erfahrung," *Holzwege*, 105–192.
    Hegel's Concept of Experience, trans. J. Glenn Gray. New
    York: Harper & Row, 1970.
  "Nietzsches Wort 'Gott ist tot'," *Holzwege*, 193–247.
    "The Word of Nietzsche: 'Got Is Dead'," in Martin Heidegger,
    The Question Concerning Technology and Other Essays,

trans. William Lovitt, 53-112.

"Wozu Dichter?" *Holzwege*, 248=295.

"What Are Poets For?" in Martin Heidegger, *Poetry, Language, Thought*, trans. Albert Hofstadter, 91-142.

"Der Spruch des Anaximander," *Holzwege*, 296-343.

"The Anaximander Fragment," in Martin Heidegger, *Early Greek Thinking*, trans. David Farrell Krell and Frank A. Capuzzi. New York: Harper & Row, 1975, 13-58.

Nietzsche (1936-1946), 2 vols. Pfullingen: Neske, 1961.

Nietzsche, vol. I: *The Will to Power as Art*, trans. David Farrell Krell. New York: Harper & Row, 1979.

Nietzsche, vol. IV: *Nihilism*, trans. Frank A. Capuzzi and David Farrell Krell. New York: Harper & Row, 1982.

Vorträge und Aufsätze (1943-1954). Pfullingen: Neske, 1961.

The collection contains eleven essays which were translated by different people for different occasions:

"Die Frage nach der Technik," VA, 13-44.

"The Question Concerning Technology," in Martin Heidegger, *The Question Concerning Technology and Other Essays*, trans. by William Lovitt, 3-35.

"Wissenschaft und Besinnung," VA, 45-70.

"Science and Reflection," in Martin Heidegger, *The Question Concerning Technology and Other Essays*, trans. by William Lovitt, 155-182.

"Uberwindung der Metaphysik," VA 71-99.

"Overcoming Metaphysics," in Martin Heidegger, *The End of Philosophy*, trans. Joan Stambaugh. New York: Harper & Row, 1973, 84-110.

"Wer ist Nietzsches Zarathustra?," VA, 101-126.

"Who Is Nietzsche's Zarathustra?," trans. Bernd Magnus in *Review of Metaphysics*, 20 (1967), 411-431.

"Was heisst Denken?," VA, 129-143.

This lecture is practically speaking identical with the first two lectures of the lecture series with the same title; cf. below.

"Bauen, Wohnen, Denken," VA, 145-162.

"Building Dwelling Thinking," trans. Albert Hofstadter, in *Basic Writings*, 323-339.

"Das Ding," VA, 163-185.

"The Thing," in Martin Heidegger, *Poetry, Language, Thought*, trans. Albert Hofstadter, 165-186.

"...dichterisch wohnet der Mensch...", VA, 187-204.

"...Poetically Man Dwells...", in Martin Heidegger, *Poetry,*

Language, Thought, trans. Albert Hofstadter, 213-229.

"*Logos* (Heraklit, Fragment 50)," VA, 207-229.

"*Logos* (Heraclitus, Fragment B 50)," in Martin Heidegger, *Early Greek Thinking*, trans. David Farrell Krell, 59-78.

"*Moira* (Parmenides VIII, 34-41)," VA, 231-256.

"*Moira* (Parmenides VIII, 34-41)," in Martin Heidegger, *Early Greek Thinking*, trans. Frank A. Capuzzi, 79-101.

"*Aletheia* (Heraklit, Fragment 16)," VA, 257-282.

"*Aletheia* (Heraclitus, Fragment B 16), in Martin Heidegger, *Early Greek Thinking*, trans. Frank A. Capuzzi, 102-123.

Platons Lehre von der Wahrheit (1942). *Mit einem Brief über den "Humanismus"* (1946). Bern: Francke, 1947.

"Plato's Doctrine of Truth," trans. John Barlow, in *Philosophy in the Twentieth Century II*, W. Barrett *et al.*, eds. New York: Random House, 1962, 251-270.

"Letter on Humanism" trans. Frank A. Capuzzi and J. Glenn Gray, in *Basic Writings*, 193-242.

Was heisst Denken? (1951-1952). Tübingen: Niemeyer, 1954.

What is Called Thinking? trans. Fred D. Wieck and J. Glenn Gray. New York: Harper & Row, 1968.

Identität und Differenz (1957). Pfullingen: Neske, 1957.

Identity and Difference, trans. Joan Stambaugh. New York: Harper & Row, 1969.

Zur Sache des Denkens. Tübingen: Niemeyer, 1969.

On Time and Being, trans. Joan Stambaugh. New York: Harper & Row, 1972.

Unterwegs zur Sprache (1950-1959). Pfullingen: Neske, 1957.

On the Way to Language, trans. Peter D. Hertz and Joan Stambaugh. New York: Harper & Row, 1966.

The first essay of US, namely "Die Sprache," was translated by Albert Hofstadter and appeared in *Poetry, Language, Thought*, 189-210.

Gelassenheit (1962). Pfullingen: Neske, 1959.

Discourse on Thinking, trans. John M. Anderson and E. Hans Freund. New York: Harper & Row, 1966.

Die Technik und die Kehre (1962). Pfullingen: Neske, 1962.

"The Question Concerning Technology," trans. William Lovitt, in *The Question Concerning Technology and Other Essays*, 3-35.

"The Turning," trans. William Lovitt, in *The Question*, 36-49.

Wegmarken (1967). Frankfurt: Klostermann, 1978.

Die Grundprobleme der Phänomenologie (1927), ed. F.-W. von Hermann. Frankfurt: Klostermann, 1975.

The Basic Problems of Phenomenology, trans. Albert Hofstadter. Bloomington: Indiana University Press, 1982.

Logik. Die Frage nach der Wahrheit (1925-1926), ed. Walter Biemel. Frankfurt: Klostermann, 1976.

Prolegomena zur Geschichte des Zeitbegriffs (1925), ed. Petra Jaeger. Frankfurt: Klostermann, 1976.

History of the Concept of Time: Prolegomena, trans. by Theodore J. Kisiel. Bloomington: Indiana University Press, 1985.

Vom Wesen der menschlichen Freiheit. Einleitung in die Philosophie (1930), ed. by Hartmut Tietjen. Frankfurt: Klostermann, 1982.

II. BOOKS AND ARTICLES ON HEIDEGGER'S PHILOSOPHY

Acquila, Richard, "Husserl and Frege on Meaning," *Journal of the History of Philosophy*, 12(1974), 377-383.

Beaufret, Jean, *Dialogue avec Heidegger*, 3 vols. Paris: Minuit, 1973-1974.

Beaufret, Jean, "Heidegger et le problème de la vérité," in *Fontaine,* 63(1947), 758-785.

Becker, Oskar, "Mathematische Existenz," *Jahrbuch für Philosophie und phänomenologische Forschung,* 8(1927).

Bergmann, Albert, "Heidegger and Symbolic Logic," in *Heidegger and the Quest for Truth*, ed. Manfred S. Frings. Chicago: Quadrangle Books, 1968, 139-162.

Biemel, Walter, *Martin Heidegger in Selbstzeugnissen und Bilddokumenten*. Reinbeck: Rohwolt, 1973. *Martin Heidegger: An Illustrated Study*. Trans. J. L. Mehta. New York: Harcourt, Brace, Jovanovich, 1976.

Birault, Henri, "Existence et vérité d'asprès Heidegger," *Revue de Métaphysique et de Morale,* 56(1950), 35-87.

Birault, Henri, *Heidegger et l'expérience de la pensée*. Paris: Gallimard, 1978.

Brentano, Franz, *On the Several Senses of Being in Aristotle*. Trans. Rolf George. Berkeley: University of California Press, 1975.

Bretschneider, Willy, *Sein und Wahrheit. Ueber die Zusammengehörigkeit von Sein und Wahrheit im Denken Martin Heideggers* (Monographien zur philosophischen Forschung, 37), (Meisenheim: Hain, 1965).

Chapelle, A., *L'Ontologie phénoménologique de Heidegger*. Paris: Editions Universitaires, 1962.

Caputo, John D., "Language, Logic, and Time," *Research in Phenomenology,* 3(1973), 147-155.

Caputo, John D., "Phenomenology, Mysticism, and the 'Grammatica Speculativa': A Study of Heidegger's 'Habilitationsschrift'," *The Journal of the British Society for Phenomenology,* 5(1974), 101-117.

Derrida, Jacques, *Margins of Philosophy.* Trans. Alan Bass. Chicago: University Press, 1982.

Düsing, K., "Objektive und subjektive Zeit. Untersuchungen zu Kants Zeittheorie und zu ihrer modernen kritischen Rezeption," in *Kantstudien,* 71(1980), 1-34.

Elliston, Frederick, ed. *Heidegger's Existential Analytic.* The Hague: Mouton, 1978.

Fay, Thomas A., "Heidegger on Logic: A Genetic Study of His Thoughts on Logic," *Journal of the History of Philosophy,* 12 (1974), 77-94.

Føllesdal, Dagfinn, "Husserl's Notion of Noema," *Journal of Philosophy,* 66(1969), 680-687.

Gadamer, Hans-Georg, "Concerning Empty and Ful-Filled Time," in *Martin Heidegger: In Europe and America,* edited by Edward G. Ballard and Charles E. Scott. The Hague: Nijhoff, 1973, 77-89.

Gadamer, Hans-Georg, *Kleine Schriften,* Vol. I: *Philosophie, Hermeneutik.* Tübingen: Mohr, 1967.

Gadamer, Hans-Georg, *Philosophische Lehrjahre: Eine Rückschau.* Frankfurt: Kostermann, 1977.

Gadamer, Hans-Georg, *Wahrheit und Methode: Grundzüge einer philosophischen Hermeneutik.* Tübingen: Mohr, 1975[4]. *Truth and Method.* New York: Seabury, Press, 1975.

Gelven, M., *A Commentary on Heidegger's "Being and Time."* New York: Harper & Row, 1970.

Gethmann, Carl F., *Verstehen und Auslegung. Das Methodenproblem in der Philosophie Martin Heideggers.* Bonn: Bouvier, 1974.

Gethmann, Carl F., "Zu Heideggers Wahrheitsfrage," *Kantstudien,* 6(1974), 186-200.

Good, P., Ed., *Max Scheler im Gegenwartsgeschehen der Philosophie.* Bern: Francke Verlag, 1975.

Grabmann, Martin, "Thomas von Erfurt und die Sprachlogik des mittelalterlichen Aristotelismus," *Sitzungsberichte der Bayerischen Akademie der Wissenschaften,* Munich, 1943.

Guilead, Reuben, *Être et liberté: Une étude sur le dernier Heidegger.* Louvain: Nauwelaerts, 1965.

Gumppenberg, Rudolph, "Die transzendentalphilosophische Urteils- und Bedeutungsproblematik in M. Heideggers 'Frühe Schriften'." *Akten des 4. Internationalen Kant-Kongresses,* Mainz, Teil II, 2(1974), 775-761.

Heinz, Marion, *Zeitlichkeit und Temporalität im Frühwerk Martin Heideggers.* Würzburg: Königshausen & Neumann, 1982.

Herrmann, Friedrich Wilhelm von, "Die Edition der Vorlesungen Heideggers in seiner Gesamtausgabe letzter Hand," *Freiburger Universitätsblätter,* 78(1982), December.

Herrmann, Friedrich Wilhelm von, *Die Selbstinterpretation Martin Heideggers.* Meisenheim am Glan: A. Hain, 1964.

Herrmann, Friedrich Wilhelm von, "Nachwort zu 'Die Grundprobleme der Phänomenologie'," Martin Heidegger *Gesamtausgabe,* Vol. 24. Frankfurt: Klostermann, 1977, 471-473.

Herrmann, Friedrich Wilhelm von, "Nachwort zu 'Sein und Zeit'," in *Sein und Zeit,* Martin Heidegger *Gesamtausgabe,* Vol. 2. Frankfurt: Klostermann, 1977, 579-583.

Herrmann, Friedrich Wilhelm von, "Zeitlichkeit des Daseins und Zeit des Seins. Grundsätzliches zu Heideggers Zeit-Analysen," *Philosophische Perspektiven,* 4(1972), 198-210.

Hilmy, S. Stephen, "The Scope of Husserl's Notion of Horizon," in *The Modern Schoolman,* 59(1981), 21-48.

Hobe, Konrad, "Zwischen Rickert und Heidegger: Versuch eines Perspektiven des Denkens von Emil Lask," *Philosophisches Jahrbuch,* 78(1971), 360-376.

Holmes, Richard, "An Explication of Husserl's Theory of the Noema," *Research in Phenomenology,* 5(1975), 143-153.

Husserl, Edmund, *Analysen zur Passiven Synthesis. Husserliana,* Vol. XI, Ed. M. Fleischer. The Hague: Nijhoff, 1966.

Husserl, Edmund, *Cartesianische Meditationen. Husserliana,* Vol. I, ed. S. Strasser. The Hague: Nijhoff, 1973[2]. *Cartesian Meditations,* trans. Dorion Cairns. The Hague: Nijhoff, 1960.

Husserl, Edmund, *Die Krisis der europäischen Wissenschaften und die transzendentale Phänomenologie: Eine Einleitung in die phänomenologische Philosophie.* The Hague: Nijhoff, 1962. *The Crisis of European Sciences and Transcendental Phenomenology,* trans. David Carr. Evanston: Northwestern University Press, 1970.

Husserl, Edmund, *Erste Philosophie,* 2 Vols. *Husserliana,* Vols. VII & VIII, ed. R. Boehm. The Hague: Nijhoff, 1959.

Husserl, Edmund, *Ideen zu einer reinen Phänomenologie und phänomenologische Philosophie. Ertes Buch: Allgemeine Einführung in die reine Phänomenologie,* ed. Karl Schuhmann. The Hague:

Nijhoff, 1976. *Ideas Pertaining to a Pure Phenomenology and to
a Phenomenological Philosophy*. First Book: *General Introduction
to Pure Phenomenology*, trans. Fred Kersten. The Hague:
Nijhoff, 1982. *Idées directrices pour une phénoménologie*, ed.
and trans. Paul Ricoeur. Paris: Gallimard, 1950.

Husserl, Edmund, *Logische Untersuchungen*, 3 vols. Tübingen:
Niemeyer, 1968; *Logical Investigations*, 2 vols., trans. J.
Findlay. New York: Humanities Press, 1970.

Husserl, Edmund, "Philosophy as a Strict Science," in *Phenomenology and the Crisis of Philosophy*, trans. Quentin Lauer. New
York: Harper Torch Books, 1965.

Janke, W., "Die reine Möglichkeit der Zukunft. Zur temporalen
Interpretation von Willen und Existenz," in W. Beierwaltes and
W. Schrader, eds., *Weltaspekte der Philosophie*. Festschrift f.
Berlinger. Amsterdam: Rodopi, 1972, 143-159.

King, Magda, *Heidegger's Philosophy: A Guide to His Basic
Thought*. New York: Dell, 1964.

Kisiel, Theodore J., "Der Zeitbegriff beim früheren Heidegger,"
in *Zeit und Zeitlichkeit bei Husserl und Heidegger, Phänomenologische Forschungen*, 14(1983), 192-211.

Kisiel, Theodore J., "On the Way to *Being and Time:* Introduction
to the Translation of Heidegger's *Prolegomena zur Geschichte
des Zeitbegriffs,*" *Research in Phenomenology*, 15(1985).

Kockelmans, Joseph J., "Destructive Retrieve and Hermeneutic
Phenomenology in 'Being and Time'," in *Research in Phenomenology*, 7(1977), 106-137.

Kockelmans, Joseph J., "Heidegger on the Self and on Kant's
Conception of the Ego," in F. Elliston, ed., *Heidegger's
Existential Analytic*, 134-156.

Kockelmans, Joseph J., "Heidegger on Time and Being," *Martin
Heidegger: In Europe and America*, edited by Edward G.
Ballard and Charles E. Scott. The Hague: Nijhoff, 1973, 55-67.

Kockelmans, Joseph J., *Martin Heidegger: A First Introduction to
His Philosophy*. Pittsburgh: Duquesne University Press, 1965.

Krell, David Farrell, "On the Manifold Meaning of *Aletheia:*
Brentano, Aristotle, Heidegger," *Research in Phenomenology*,
5(1975), 77-94.

Kuhn, Helmut, "The Phenomenological Concept of Horizon," in
Philosophical Essays in Memory of Edmund Husserl, ed. Marvin
Faber. Cambridge, Mass.: Harvard University Press, 1940, 106-
123.

Langan, Thomas, *The Meaning of Heidegger: A Critical Study of
Existentialist Phenomenology*. New York: Columbia University

Press, 1959.

Lask, E., *Die Lehre vom Urteil, Gesammelte Schriften,* vol. II, ed. Eugen Herrigel. Tübingen: Mohr, 1923.

Lask, E., *Die Logik der Philosophie und die Kategorienlehre: Eine Studie uber den Herrschaftsbereich der logischen Form.* Tübingen: Mohr, 1911.

Lehmann, Karl, "Metaphysik, Transzendentalphilosophie und Phänomenologie in den ersten Schriften Martin Heideggers (1912-1916)," *Philosophisches Jahrbuch,* 71(1964), 333-367.

Macquarrie, John, *Martin Heidegger.* Richmond: John Knox Press, 1968.

Mohanty, J. N., "On Husserl's Theory of Meaning," *Southwest Journal of Philosophy,* 5(1974), 229-243.

Morscher, Edgar, "Von der Frage nach dem Sein von Sinn zur Frage nach dem Sinn von Sein--der Denkweg des frühen Heidegger," *Philosophisches Jahrbuch der Görres Gesellschaft,* 80(1973), 379-385.

Müller-Lauter, W., *Möglichkeit und Wirklichkeit bei Martin Heidegger.* Berlin: de Gruyter, 1960.

Petzet, Heinrich Wiegand, *Auf einen Stern zugehen: Begegnungen und Gespräche mit Martin Heidegger, 1929-1976.* Frankfurt: Societäts-Verlag, 1983.

Peursen, Cornelius van, "The Horizon," in *Husserl: Expositions and Appraisals,* ed. P. Elliston and P. McCormick. Notre Dame: Notre Dame University Press, 1977, 182-201.

Pietersma, Henry, "The Concept of Horizon," *Analecta Husserliana,* 2(1972), 278-282.

Pöggeler, Otto, "Den Führer führen? Heidegger und kein Ende," *Philosophische Rundschau,* 32(1985), 26-67.

Pöggeler, Otto, *Der Denkweg Martin Heideggers.* Pfullingen: Neske, 1963, 1983².

Pöggeler, Otto, ed. *Heidegger: Perspektiven zur Deutung seines Werkes.* Köln: Kiepenheuer & Witsch, 1969.

Pöggeler, Otto, "Heidegger und das Problem der Zeit," in *L'Héritage de Kant. Festschrift* f. M. Régnier. Paris: Beauchesne, 1982.

Pöggeler, Otto, "Zeit und Sein bei Heidegger," in *Zeit und Zeitlichkeit bei Husserl und Heidegger. Phänomenologische Forschungen,* 14(1983), 152-191.

Richardson, William J., "Heidegger's Way Through Phenomenology to the Thinking of Being," in Thomas Sheehan, ed., *Heidegger: The Man and the Thinker,* 79-93.

Richardson, William J., *Heidegger: Through Phenomenology to*

Thought. The Hague: Nijhoff, 1963.

Rickert, Heinrich, "Das Eine, die Einheit und die Eins. Bemerk-ungen zur Logik des Zahlbegriffs," *Logos,* 2(1911), 26-78.

Rickert, Heinrich, *Gegenstand der Erkenntnis.* Tübingen: Mohr, 1921.

Rickert, Heinrich, *Grenzen der naturwissenschaftlichen Begriffsbildung.* Tübingen: Mohr, 1931.

Ricoeur, Paul, "Phenomenology and Hermeneutics," in *Hermeneutics and the Human Sciences,* ed. and trans. by John Thompson. Cambridge: University Press, 1981, 101-128.

Rosales, A., "Observaciones criticas a la idea de temporalidad propria en 'Ser y Tiempo' de Heidegger," in *Revista Venezolana de Filosofia,* 8(1978), 83-96.

Rosales, A., *Transzendenz und Differenz. Ein Beitrag zum Problem der ontologischen Differenz beim frühen Heidegger. Phaenomenologica,* vol. 33. The Hague: Nijhoff, 1970.

Schelling, Friedrich Wilhelm Joseph, *Philosophie der Offenbarung,* in *Sämmtliche Werke,* vol. II, 3. Stuttgart: Cotta, 1856-1861.

Schmitt, Richard, *Martin Heidegger on Being Human: An Introduction to "Sein und Zeit."* New York: Random House, 1969.

Seebohn, Thomas M., *Zur Kirtik der hermeneutischen Vernunft.* Bonn: Bouvier, 1972.

Sheehan, Thomas, ed., *Heidegger: The Man and the Thinker.* Chicago: Precedent Publishing Inc., 1981.

Sheehan, Thomas, "Heidegger's Early Years: Fragment for a Philosophical Biography," in *Heidegger: The Man and the Thinker,* 3-19.

Sheehan, Thomas, "The 'Original Form' of *Sein und Zeit:* Heidegger's *Der Begriff der Zeit* (1924)," *Journal of the British Society for Phenomenology,* 10(1979), 78-83.

Sherover, Charles, *Heidegger, Kant, and Time.* Bloomington: Indiana University Press, 1971.

Sinn, Dieter, "Heidegger's Spätphilosophie," in *Philosophische Rundschau,* 14(1967), 81-182.

Sokolowski, R., *Husserlian Meditations.* Evanston, Ill.: Northwestern University Press, 1974.

Stewart, R.M., *Psychologism, Sinn and Urteil in the early Writings of Heidegger.* Unpublished Dissertation, Syracuse University, 1977.

Stewart, R.M., "The Problem of Logical Psychologism for Husserl and the Early Heidegger," *The Journal of the British Society for Phenomenology,* 10(1979), 184-193.

Tugendhat, Ernst, *Der Wahrheitsbegriff bei Husserl und Heidegger*. Berlin: de Gruyter, 1967.

Tugendhat, Ernst, *Selbstbewusstsein und Selbstbestimmung*. Frankfurt: Suhrkamp, 1979.

Watson, James R., "Heidegger's Hermeneutic Phenomenology," *Philosophy Today,* 15(1971), 30-43.

Welch, Cyril, "Review of *Frühe Schriften* by Martin Heidegger," *Man and World,* 7(1974), 87-91.

III. COLLECTIONS OF ESSAYS ON HEIDEGGER'S THOUGHT

Ballard, Edward G. and Charles E. Scott, eds. *Martin Heidegger: In Europe and America*. The Hague: Nijhoff, 1973.

Frings, Manfred S., ed. *Heidegger and the Quest for Truth*. Chicago: Quadrangle Books, 1968.

Gadamer, H. G., W. Marx, C. F. von Weizsäcker, eds. *Heidegger: Freiburger Universitätsvorträge zu seinem Gedenken*. Freiburg: Alber, 1977.

Guzzoni, Ute, ed. *Nachdenken über Heidegger*. Hildesheim: Gerstenberg Verlag, 1980.

Kockelmans, Joseph J., ed. *On Heidegger and Language*. Evanston: Northwestern University Press, 1972.

Murray, Michael, ed. *Heidegger and Modern Philosophy: Critical Essays*. New Haven: Yale University Press, 1978.

Pöggeler, Otto, ed. *Heidegger: Perspektiven zur Deutung seines Werkes*. Cologne: Kiepenheuer und Witsch, 1969.

Sallis, John C., ed. *Heidegger and the Path of Thinking*. Pittsburgh: Duquesne University Press, 1970.

Sheehan, Thomas, ed. *Heidegger: The Man and the Thinker*. Chicago: Precedent Publishing, Inc., 1981.

Anteile. Martin Heidegger zum 60. Geburtstag. Frankfurt: Klostermann, 1970.

Durchblicke. Martin Heidegger zum 80. Geburtstag. Frankfurt: Klostermann, 1970.

Martin Heidegger zum siebzigsten Geburstag. Pfullingen: Neske, 1959.

INDEX OF NAMES

INDEX OF SUBJECTS

truth, 146, 147; and truth and freedom inseparable, 163; articulation of Being a basic problem of temporal ontology, 197; as *aletheia*, 175, 177; as *Anwesenheit*, 157; as a process of non-concealment, 163; as Being-discovered, 156; as Being-true, 156; as "category," 83; as continuous and constant coming-to-presence, 153, 155, 156; as continuous present, 235, 236; as continuous presence in metaphysics, 232, 233, 236; as convertible to a continuously present being, 237; as *energeia*, 157; as event of appropriation *(Ereignis)*, 234; as history, 178; as *Idea*, 236; as *Logos*, 177, 178; as *Nomos*, 177, 178; as originary phenomenon, 83; as *ousia*, 155, 156, 157; as permanent presence, 134; as representation, 236; as representativeness in modern metaphysics, 235; as that which remains unthought, 95; as the Free, 175, 176, 177; as the horizon of the world, 165, 170; as the issue of Heidegger's thought, 94; as the lighting process, 165; as the original moral law, 177; as the process of *ale-theia*, 173; as the supremely Free, 175; as the *transcendens* pure and simple, 232; as the unavailable event of appropriation, 239; as what is con-

tinuously present, 237–238, 242; as what-is-true *(alethes)*, 157; as will to power in Nietzsche, 236; as will to power willing itself eternally, 236; e-mits itself, 173; excessiveness of, 83–85, 87; founded in a divine being, 236; has priority over Dasein, 173; in its own truth the unavailable history, 234; inquiry into its truth is a regress into the hidden ground of metaphysics, 234; is Being of beings, 137; its truth-character a basic problem of temporal ontology, 197; its sense is ground for the diversification in the multiplicity of the meaning of Being, 231; its truth as the ground of metaphysics, 234; its truth as appropriation is the unavailable history, 235, 240; its truth cannot be called ground, 243; its truth is to be articulated as the non-ground of the event of appropriation, 244; its truth is what is left unthought in metaphysics, 245; its unity is a basic problem for temporal ontology, 197; lies beyond the transcendental modes of Being, 232; liberates man unto the Free, 175; multiplicity and unity of the kinds of its meaning, 231; multiplicity of its meaning, 229; not a being, 173; not a real predicate, 197; not the ulti-

Modus significandi, 4, 7, 9, 19; *activus*, 4, 5, 19, 20; *passivus*, 4, 5, 6
Moods, 211
Morality, and freedom in early Heidegger, 171; in EM, 182; in the later Heidegger, 176; new morality, 177
Moral Ought, 172
Motivation, 111-112
Movement, 46
Mystery, and errance, 173-174; as the concealment-revealment of Being, 173; *See* Non-truth

"Natural" Law, 172
Natural Standpoint, 50
Naught, as belonging to Being, 239; as veil of Being, 239; of Being not the absolute nothing, 234
Nihilism, 239
Noema, 6, 48
Noematic Sense, 3, 4, 5, 6
Noesis, 48
Noetic Act, 3
Nomos, as dispensation of Being, 177, 178; as original moral law, 177, 178
Non-being *(Nichts)*, 167
Non-Being, different ways of, 154
Non-truth, as mystery and errance *(Irre)*, 174; is dissimilation, 153; is not hiddenness, 153
Now-Time, 199

Objectification, and nature of thought, 231
Ontic Meaning, 13
Ontological Difference, 84, 88, 150, 235; as basic problem of temporal ontology, 197; as difference between Being and beings, 232-233; categorial vs. transcendental, 158
Ontological Disposition *(Befindlichkeit,* mood), v, 166
Ontological Interpretation, ix, 117
Ontological Truth, as non-concealment, 153. *See* Truth.
Ontology, viii, 29, 32, 94, 96, 127-128; and fundamental ontology, 232; and the objectivation of Being, 194; as a non-deductive genealogy of the different ways of Being, 184; as inquiry into the Being of beings, 232; as the interpretation of the prescientific comprehension of Being, 194; as the objectivation of the preontological understanding of Being, 196; destruction of, 127-143; method of, 127; of Kant is a regional ontology of the temporal ontology, 206; regional, 2, 9; temporal vs. fundamental, 194; time is its central theme, 184
Origin, as historical, 91-97; as original issue, 91, 97-100; as radical origin of philosophical thought, 91, 101-103; threefold meaning of term, xiii-xiv

Parousia, 55, 56, 57; temporality of, 57